Geopolitics in Antarctica

Polar Research Series

edited by Bernard Stonehouse

L.W. Brigham, **The Soviet Maritime Arctic**

C. Harris and B. Stonehouse, **Antarctica and Global Climatic Change**

R.D. Massom, **Satellite Remote Sensing of Polar Regions**

M. Nuttall, **Arctic Homeland**

H.P. Huntington, **Wildlife Management and Subsistence Hunting in Alaska**

B.A. Riffenburgh, **The Myth of the Explorer**

S. Chaturvedi, **The Polar Regions: A Political Geography**

K. Dodds, **Geopolitics in Antarctica**

Geopolitics in Antarctica

Views from the Southern Oceanic Rim

Klaus Dodds
Royal Holloway, University of London

Published in association with the

Scott Polar Research Institute
University of Cambridge

by

JOHN WILEY & SONS
Chichester • New York • Weinheim • Brisbane • Singapore • Toronto

National 01243 779777
International (+44) 1243 779777
e-mail (for orders and customer service enquiries):
cs-books@wiley.co.uk
Visit our Home Page on http://www.wiley.co.uk
or http://www.wiley.com

Other Wiley Editorial Offices

John Wiley & Sons, Inc., 605 Third Avenue,
New York, NY 10158-0012, USA

WILEY-VCH Verlag GmbH, Pappelallee 3,
D-69469 Weinheim, Germany

Jacaranda Wiley Ltd, 33 Park Road, Milton,
Queensland 4064, Australia

John Wiley & Sons (Asia) Pte Ltd, 2 Clementi Loop #02-01,
Jin Xing Distripark, Singapore 129809

John Wiley & Sons (Canada) Ltd, 22 Worcester Road,
Rexdale, Ontario M9W 1LI, Canada

Library of Congress Cataloging-in-Publication Data

Dodds, Klaus.
Geopolitics in Antarctica : views from the Southern Oceanic Rim / Klaus Dodds.
p. cm. — (Polar research series)
Includes bibliographical references and index.
ISBN 0-471-96992-3 (alk. paper)
1. Geopolitics—Antarctica. I. Title. II. Series.
G877.D63 1997
327.8—dc21 97–14902
CIP

British Library Cataloguing in Publication Data
A catalogue record for this book is available from the British Library

ISBN 0-471-96992-3

Typeset in 10/12pt Times from the author's disks by Dorwyn Ltd, Rowlands Castle, Hants.
Printed and bound in Great Britain by Biddles Ltd, Guildford and King's Lynn
This book is printed on acid-free paper responsibly manufactured from sustainable forestation, for which at least two trees are planted for each one used for paper production.

For John Dodds (1934–1988)

Contents

List of figures

Preface

The American novelist, Barry Lopez, once noted that the Antarctic was:

> The only continent never to have known war, [the Antarctic] lies immense in the middle of the Southern Ocean. No horse stands to pasture here, no automobile waits at the light, no tree huffs before a close wind . . . It lies like a medieval keep, bunkered behind an apron of sea ice as broad as Europe, behind walls of jade, chalk and turquoise ice that falls starkly to the frozen ocean. Inland, but for a few bare valleys hardly distinguishable from the rainless deserts of Mars, stretches a landscape so unreachable, so indifferent, it seems autistic (Lopez 1988).

In the last 10 or 15 years, the pristine geography of the Antarctic and the Southern Ocean has become increasingly accessible to the policy maker, the scientist and the tourist. The key turning point in the political transformation of the polar continent came during the negotiations over a possible minerals regime in the 1980s. At these meetings conducted by the Consultative Parties of the Antarctic Treaty System (ATS), the Antarctic became a source of considerable debate and controversy as, on the one hand, members of the United Nations General Assembly debated the merits of declaring the continent part of the common heritage of mankind and, on the other hand, environmental movements such as Greenpeace called for the complete banning of mining and mineral exploitation (Parsons 1987, Herber 1991, Suter 1991). The eventual rejection of the Convention on the Regulation of Antarctic Mineral Resource Activities (CRAMRA), after six years of formal negotiations, was the political low point for the ATS, which had effectively governed the Antarctic since 1961 (Brown 1995, Wolfrum 1995, Chaturvedi 1996). Opposition to CRAMRA later proved irresistible as it was argued by ecological movements such as Friends of the Earth and Greenpeace that the establishment of exploitative rights under

CRAMRA placed the Antarctic continent in environmental peril (see Suter 1991, Davis 1992a, Bush 1995, Joyner 1995, Dalziell 1997).

In the aftermath of CRAMRA and the Cold War, the Consultative Parties of the ATS successfully negotiated the Protocol on Environmental Protection (Protocol) in 1991. This is the most important legal convention ever to be applied to the Antarctic and surrounding oceans. It is hoped that by the 1997 Consultative Party meeting in Christchurch, the Protocol will have been ratified by all the Consultative Parties attached to the ATS (Dalziell 1997). The creation of a Committee on Environmental Protection (CEP) will have the difficult task of implementing comprehensive ecosystem management for the continent. Whilst the immediate threat of mining has been removed, considerable problems remain that have yet to be resolved. Five major issues can be identified at this stage (see Rothwell 1992). First, since the application of the Protocol will have to be related to existing conventions and agreed measures dealing with living resource management and environmental protection, it is not, therefore, being carried out in a legal and political vacuum (van Bennekom 1992). Second, the mining issue and the controversy over legal and property rights in the Antarctic and Southern Ocean needs to be resolved because the ban on mining is temporary and relies on the self-restraint of states. Third, the role of states not party to the ATS and the 1959 Antarctic Treaty remains uncertain. The case of Pakistan is the most obvious and pressing given that it launched a number of expeditions to the Antarctic in the 1990s but has not, to date, joined the ATS (Rabbani 1994). Fourth, the political and legal enforcement of the Protocol remains uncertain regardless of whether a particular party is a claimant or non-claimant state. At the heart of the issue is territorial sovereignty. The French decision to build a runway in Adélie Land during 1987 in spite of considerable environmental and political condemnation of their destruction of penguin nesting areas, is instructive here. The construction work continued because the incumbent government argued that it was exercising its sovereign rights on French territory. Finally, the effective regulation of tourism raises once again the sensitive issues of environmental liability, legal jurisdiction and territorial sovereignty (see Beck 1994a, Chaturvedi 1994). Some countries, in particular Chile and New Zealand, have urged that the CEP address the issue of responsibility for management and regulation with regard to the complex liabilities relating to tourist ships and visitors (Vidas 1992, Johnston 1997). The Protocol has approached some of these issues but there remains a considerable amount of unfinished legal and political business.

These are the immediate circumstances facing the major southern hemispheric states in respect of their participation in the Antarctic and the Southern Ocean. Without an appreciation of the historical and political context of their involvement in the region, it is difficult to comprehend how states such as Argentina, Australia and South Africa respond to the recent challenges posed by the Protocol (see Beck 1986, Chaturvedi 1990). In contrast to their northern hemispheric counterparts, the geographical proximity of the Antarctic has exercised a powerful hold on the collective and individual geographical imaginations of these southern hemispheric nations (see Child 1988). In this sense, the Antarctic cannot be represented as having an inert or 'frozen' geography. The Antarctic as a place has been, among other things, a dynamic marker of national identity and prestige. For all these nations, the Antarctic has occupied a fundamental part of their national histories and in some cases is even considered part of their national territory (see Pinochet de la Barra 1955, Candioti 1961). It has been argued that: "The Antarctic holds special strategic value to the nations of South America – particularly Argentina and Chile . . . The national sentiments run especially high in Argentina and Chile, whose territory lies closest to the cold continent, separated by 450 miles of frigid seas across the Drake Passage" (Joyner 1985:2).

A concern for their Antarctic claims has also been related to a wider consideration of national security and regional priorities. The collective endeavours of the southern hemispheric nations to declare the Antarctic a zone of peace and co-operation in the 1950s was indicative of this concern for a geographically proximate place. The New Zealand government has been a tireless advocate of linking southern hemispheric nuclear free zones. As their White Paper on the Antarctic Environment in 1989 noted: "Our fundamental priority is that it [Antarctica] remains peaceful and nuclear free. The Antarctic Treaty is a superb disarmament model for the world. It established the world's first nuclear free zone . . . We have an effective guarantee that, to our south, peace and security are assured" (New Zealand 1989). More recently, at the August 1996 South Pacific Forum meeting, New Zealand Prime Minister Jim Bolger unveiled his proposals for a Southern Hemispheric Nuclear Free Zone which would effectively link existing geographical zones of peace covering the South Atlantic, the Indian Ocean, the South Pacific and the Antarctic.

This book is concerned with the geopolitics of the Antarctic and the Southern Oceanic Rim States (SORS). The term SORS was chosen in order to identify the states that are geographically proximate to the Antarctic and the Southern Ocean rather than southern hemispheric states *en masse*. Within human geography and development studies, there have been a considerable number of critiques placed against arbitrary labels such as the "Developing World" and or "Third World" (see Wolf-Philips 1987, Escobar 1995). The designation is intended, therefore, to be explicitly geographical in the sense of location and physical proximity. Furthermore, the selection of nations under consideration here was based on geographical, political and practical considerations. The choice can be judged in terms of their political commitment to the ATS; their historic involvement in the region; scientific investment in the region and the claims they press either to the polar continent and or island groups in the Southern Ocean. Four states claim a portion of the Antarctic continent (Argentina, Australia, Chile and New Zealand) whilst the other two have rejected these and other polar territorial claims (South Africa and India). The Australasian states have been classified as industrialized and/or developed, whereas the South American states are usually considered not only as developing nations but also as the most hawkish in the defence of their Antarctic territories (see Pinochet de la Barra 1955, Child 1988, Escudé 1988). The polar interests of the three old Commonwealth states have been closely defined through their earlier imperial connections and subsequently through close relations with the United Kingdom in the Antarctic Treaty System (Swan 1961, Beck 1986, Dodds 1996a). The decision to include India, unlike Brazil and Uruguay, in this analysis was taken on the geographical grounds that its location is directly proximate to the Southern Ocean and on the basis of its political importance to the international relations of the Antarctic since the 1950s. India has been an active participant, since the 1970s, in terms of its resource-related and scientific endeavours within the Indian Ocean and the Antarctic continent (Chaturvedi 1990, Dey 1992). Moreover, India's role in the ATS has been interesting in the sense of exposing some political tensions between the SORS and other southern hemispheric and/or developing nations. However, Brazil's interests in the region do feature in this account when it relates to South American co-operation in the South Atlantic and the Antarctic (see De Castro 1976, Azambuja 1981, Child 1988).

In attempting to explain the political development and historical evolution of polar involvement of the SORS, four major questions will be tackled: Why did some of those countries participate in the formal annexation of the Antarctic and others not? Why did the SORS participate in the discussions leading to the 1959 Antarctic Treaty and others not? How have the political, economic and resource interests of the SORS in the region changed over time? And, finally, how are the SORS responding to contemporary international concerns about environmental protection and the regulation of activities such as polar tourism?

This account suggests that the geographical proximity and the presumed geopolitical significance of the Antarctic occupy key explanatory positions. The geographical proximity of the Antarctic to those southern hemispheric nations played a part in stimulating voyages of exploration, territorial aggrandizement, resource exploitation and scientific investigation. The changing geographies of the polar continent and surrounding oceans are also critical in this respect. As the polar continent became subjected to different technologies of exploration, communication and travel, so representations of Antarctica altered over time and space (Dodds 1997). The geopolitical appreciation of Antarctica depended upon the collection of comprehensive geographical knowledge as well as changing political factors such as the Cold War and the development of the Antarctic Treaty System (cf. Osherenko and Young 1989 on the Arctic).

The following chapters on the SORS are not exhaustive in terms of their consideration of legal, political, geographical and scientific issues affecting the polar continent. There are a sufficient number of specialist accounts of individual SORS to allow the interested reader to follow through some of these observations (Marsh 1948, Pinochet de la Barra 1955, Candioti 1960, Swan 1961, Logan 1979, Chaturvedi 1990). Recent accounts are also considerably more detailed in terms of policy development and the interaction of government departments or institutions (Infante 1993, Herr and Davis 1996). In addition there is little point in rehearsing well-established facts such as the details of the 1959 Antarctic Treaty or specific resource conventions (Beck 1986, Chaturvedi 1996, Stokke and Vidas 1996). The chapters on the Commonwealth nations of Australia, South Africa and New Zealand have been enriched by archival material collected in these countries and in others such as Canada and Britain. There is a degree of historical and political repetition in these chapters because their polar histories are so intertwined with former imperial polar policy and the activities of the United States. In contrast, the chapters on the two South American states are less archivally based and draw rather more on public documents such as academic journals, reports and public speeches. This is largely a reflection of government sensitivity attached to documents relating to polar matters and national territorial claims. India's involvement in the Antarctic, however, is comparatively recent: since the first Indian polar expedition in 1981. Historical time frames, therefore, do alter within this account.

Structure of book

The opening chapter sets out to introduce the significant themes which will permeate this book. These include: the historical and cultural importance of the Antarctic to the SORS, the role of the SORS as middle powers and the legal and territorial geographies of the Antarctic. The purpose of this material is to provide a broad context from which the subsequent specialist chapters can be related to one another.

The second chapter is concerned with geopolitics and explains why a critical geopolitical approach has been adopted (see Dalby 1991, Ó Tuathail and Agnew 1992, Ó Tuathail 1996). With very few exceptions such as Sanjay Chaturvedi, most authors who cite the term "geopolitics" in their work on the Antarctic have little apparent interest in theorizing about or examining the intellectual and political terrain marked by geopolitics (Chaturvedi 1996). If the term is cited it is usually intended to be indicative of either geographical location *per se* or of a hard-headed realist approach to the international politics of the Antarctic (Joyner 1985, Roucek 1981, Glassner 1995). Leaving aside the physical changes to the Antarctic continent, recent research in human geography has argued that places should be considered as being socially produced and, as a consequence, capable of changing in significance (Agnew 1993, Agnew and Corbridge 1995). There is no reason, therefore, to assume that the Drake's Passage is of great geopolitical importance just because it happens physically to divide the South American mainland from the Antarctic. Strategic significance is determined usually on the basis that certain parts of the world are encoded with strategic value at the expense of others. Historical considerations have influenced the significance of the Antarctic in the present century in so far as the physical shape of the continent has only fairly recently been completely mapped. Geopolitics as a form of geographical reasoning rather than as a rigid policy mindset can indicate how political, scientific and media elites "manufacture" a place and then use that construction to justify or legitimate their polar policies or claims to territory.

The main section of this book is devoted to the six SORS: Argentina, Australia, Chile, India, New Zealand and South Africa. There is a review of past episodes of exploration and an explanation of national polar claims. The major part of each chapter is devoted to explaining how geopolitical reasoning has been employed to explain why these states are interested in the Antarctic region and how these concerns were located in a wider regional setting. Interest in the Antarctic was, therefore, always related to geopolitical concerns such as regional security pacts in the South Atlantic, resource management and other issues affecting the Southern Ocean (Dodds 1994a). Four out of the six states were regional supporters of the American-led security and economic coalition during the Cold War. Argentina and Chile were part of the Inter-American security system, and Australia and New Zealand were part of ANZUS and SEATO (Child 1985, Yahuda 1996). South Africa wished to participate in a Western security system but failed, because of apartheid, which condemned her to a pariah status (see Barber and Barratt 1990, Geldenhuys 1990), whilst India's alignment alternated between both superpowers during the Cold War (see Hewitt 1992). The ATS was remarkably successful in preventing the Antarctic from becoming a scene of Cold War confrontation but it was neither immune from the ideological reasoning of the Cold War nor considered irredeemably removed from that global confrontation (cf. Beck 1990c, Mason 1991). There was no reason to presume in the 1960s (especially after the 1962 Cuban Missile Crisis) that the ATS would have survived and even prospered for 30 or more years.

The final chapter is devoted to the future management of the Antarctic and the implications for the SORS. This group of nations will be fundamental in shaping the future political and environmental management of the region. With the creation of the Valdivia Group (1995), southern hemispheric states have consolidated their regional relationships and supplemented the existing regional networks such as the South Atlantic Zone of

Peace and Co-operation. Underlying this development is the belief that the interests of the SORS differ from those states located in the northern hemisphere. With recent admission of Brazil to the Valdivia Group in 1996–7, the SORS will not only have to continue to negotiate their relationships with northern hemispheric states but also manage their mutual concerns with other "southern" countries.

Acknowledgements

During the course of writing this book, I have been fortunate to receive a great deal of academic and financial support from a variety of sources. I wish to thank my former teachers at the University of Bristol for their initial support, in particular Leslie Hepple and Nigel Thrift. Thereafter, colleagues at the University of Edinburgh and Royal Holloway, University of London have been an important source of inspiration and friendship. I owe special thanks to Denis Cosgrove and David Simon who commented on earlier drafts of the book. Justin Jacyno of the Department of Geography, Royal Holloway kindly carried out the cartographic work. A number of retired and or serving civil servants responsible for polar policy in Argentina, Australia, Chile, India, New Zealand, South Africa, and the United Kingdom granted me interviews. Most of the interviews were carried out on a non-attributable basis, but I acknowledge their assistance in this project. Finally, I have also been fortunate to be working within a small group of critical geopolitical writers and polar studies scholars who have greatly influenced my research on the Antarctic and Southern Ocean. Grateful thanks in particular to David Atkinson, Sanjay Chaturvedi, Simon Dalby, Beau Riffenburgh, James Sidaway, Gerard Toal and Peter Taylor. Peter Beck offered many valuable comments on the complete manuscript. However, the responsibility for any shortcomings is mine alone.

This research has resulted in a number of different field trips throughout the southern hemisphere over a four-year period. I have had the benefit of a number of research grants from the following bodies: the British Academy, the British Council, the Nuffield Foundation, Royal Holloway, University of London and the University of Bristol. My thanks to their Trustees and to sympathetic Heads of Department in Bristol, Edinburgh and Royal Holloway.

Over the last four years, I have drawn upon the services and expertise of staff attached to a large number of institutions: Archivo General de la Naçion (Buenos Aires); Antarctic Division, Department of Education, Sport and Territories (Hobart); Australian Archives (Canberra); British Antarctic Survey (Cambridge); Department of Environmental Affairs and Tourism (Pretoria); Department of Foreign Affairs and Trade (Canberra); Instituto Antartico Argentino (Buenos Aires); Instituto Antartico Chileno (Santiago de Chile); National Archives of Canada (Ottawa); National Archives of New Zealand (Wellington and Christchurch); Ministry of Foreign Affairs and Trade (Wellington); Ministry of Foreign Affairs and Worship (Buenos Aires); Scott Polar Research Institute (Cambridge); South African Archive Service (Pretoria); South African Institute for International Affairs (Johannesburg); South African Research Library (Cape Town). Special thanks to Bob Headland, Philipa Hogg and William Mills at the Scott Polar Research Institute in Cambridge for dealing with numerous requests for information and archival access over this research period. Bernard Stonehouse, in his capacity as the editor of the Polar Research Series, offered some valuable comments and suggestions during the revision process. My final thanks goes to the staff of John Wiley and Sons, Chichester and London for their help in finalizing the publication of this book.

Finally, I remain indebted to my family for their constant encouragement and financial support. Aloisius played a key role in casting a critical geopolitical eye over the final text. This book is dedicated to my late father, who as a former Royal Naval officer first fired my geographical imagination for the ports and cities of the southern hemisphere.

Klaus Dodds,
18 February 1997

1

Introduction: the Southern Oceanic Rim States and Antarctica

The purpose of this opening chapter is to set the scene for the subsequent detailed examination of the SORS and their involvement in the Antarctic and Southern Ocean. Three major themes have been identified because they contain considerable relevance for this investigation: first, the historical and political significance of the Antarctic to the southern hemispheric nations. This account suggests that their relationship to the region is distinct from northern hemispheric nations such as France, the United States or the United Kingdom. The high profile of the Antarctic and Southern Ocean within public education and the mass media is used to make that particular point. Second, the concept of middle-power diplomacy is discussed with particular reference to the SORS. The intention is not to enter the contested terrain of international political economy but rather to make the point that these nations could be described as middle powers in terms of their geographical and political relationships to the region. Third, the legal and political geographies of the Antarctic and Southern Ocean are noted with direct reference to territorial sovereignty, the Law of the Sea and the Antarctic Treaty System. The ultimate purpose of this chapter, therefore, is to introduce various themes that will permeate this account of the SORS.

Historical legacies and public relevance

In his foreword to *New Zealand and Antarctica*, the then Foreign Minister, Don McKinnon, noted that: "We [New Zealand] have had intimate links with Antarctica since we claimed sovereignty over Ross Dependency in 1923. Awareness of its remoteness and beauty is part of our national consciousness, built up over decades of exploration by New Zealanders in collaboration with other nations" (New Zealand 1996b). From the mid-1950s onwards, the Antarctic and Southern Ocean have been important elements in the political and educational programmes of successive New Zealand governments. Local New Zealand newspapers such as the *Greymouth Evening Star* and the *Te Aroha News* played their role in raising public consciousness of the region. Moreover, it was not uncommon in the 1950s for forthcoming Antarctic endeavours to give rise to musings about the colonial foundation of these southern hemispheric states. As the *Southland Times* recorded on 13 October 1955: "and the appeal to Southlanders comes at a time when we are thinking about the pioneers of this province and the hardships they overcame in making it productive and a pleasant place to live in . . . Our participation in the Commonwealth expedition [1955–8 Commonwealth Trans-Antarctic Expedition] will be a sign, after a century in this young land, of our national maturity." Likewise in other southern hemispheric nations, the geographical proximity of the Antarctic has meant that this is a part of the world that commands considerably more importance than would be accorded to it by northern hemispheric counterparts. The long-standing claims to Antarctic territory by southern hemispheric states such as New Zealand and Argentina have cemented this relationship, a fact recognized by Britain in the 1950s when it contemplated giving up its formal claims to the Falkland Islands Dependencies in the face of an overlapping Argentine claim to an Antarctic sector (Beck 1986, Dodds 1996b).

Polar endeavours, to echo the words of Edward Said, often had a cultural impact on the metropolitan landscapes of the explorers and colonizers (Said 1993). A striking feature of many southern hemispheric cities is the way urban architecture records the long-standing legacy of polar exploration, discovery and commercial endeavour. Commercial whaling and private expeditions were organized from cities such as Buenos Aires, Cape Town, Christchurch, Durban, Hobart and Punta Arenas. In Christchurch, these events and activities have been richly commemorated through public statues (Scott and Cook), a museum (Canterbury Museum) and an information centre (International Antarctic Centre) which seek to promote public awareness of New Zealand's role in the Antarctic region. The recent creation of Antarctica New Zealand in 1995 as a crown agency responsible for the New Zealand Antarctic Programme witnessed the importance attached to public education. The major strategic goals of the Agency are: "the leadership and co-ordination within New Zealand of activities promoting public and scientific awareness of the unique characteristics and importance of the Antarctic and Southern Ocean region . . . From the debating chambers of the Antarctic Treaty System to the classrooms of New Zealand we aim to be an influence" (New Zealand 1996b).

In other southern hemispheric nations such as Australia or Argentina, the official construction and reproduction of national identities has frequently drawn upon memories of past feats of polar exploration and/or contemporary events in the region. The Antarctic is often represented in maps, flags, anthems, stamps, stories and public exhibitions (Monmonier 1991, Dodds 1993). Interest in the region is also sustained through a combination of public television programmes, the print media and government-sponsored visits to the Antarctic by parties of schoolchildren or other groups of citizens (Child 1988). The geographic practices and discourses which imaginatively and physically map out the terrain of the state assist in constructing official and popular geographies of the Antarctic and the Southern Ocean. In Argentina, the Instituto Geografico Militar (IGM) has been charged with producing the official maps and surveys of Argentina (Morris 1996). Since 1948, it has been instructed to include on all national maps cartographic depictions of Antartida Argentina and Las Islas Malvinas (Dodds 1993). English names have been replaced (eg The Falkland Islands and South Sandwich Islands) with Argentine names (eg Islas Malvinas and Islas Sandwich del Sur), and British and Chilean claims over the Antarctic peninsula have been omitted from all official map production by Presidential Decrees. These "intentional silences" on the map (and then reproduced on postage stamps) are just one part of the IGM's mission (Dodds 1993). From an early age, Argentine citizens are taught that their nation is composed of three parts: mainland, insular (including the Malvinas) and the Argentine Antarctic sector. Cartographic representations of a tri-continental Argentina are to be found in road atlases, newspapers and on public mural maps. Cartography, therefore, provides a means of promoting popular geographical consciousness and as an expression of a will to order in the developmental plans of the state. The Argentine Antarctic sector has often been depicted, within official development plans, as representing a promising region for the future development of Argentina (Daus 1950:23–4).

The construction of these official cartographies reflect the considerable political power of the map and the survey (Harley 1988, 1992). State-sanctioned histories in Latin America have played an important role in "explaining" the territorial evolution of states and the shapes of their boundaries both physical and imaginative. In the Argentine case, Carlos Escudé has written extensively on the construction and reproduction of historical fictions of "territorial loss" and has argued that official representations of national history consistently suggest that Argentina has been the victim of territorial loss at the expense of neighbouring countries such as Chile and Bolivia (Escudé 1986, 1988, 1992). These stories of territorial loss (updated in the 1940s to include polar territories) have been subsequently reproduced in the formal educational system and in the print and televisual media. Ironically, the Argentine state successfully expanded northwards, southwards and westwards, encapsulating huge tracts of land throughout the nineteenth century. The only territorial "loss" was the Falklands/Malvinas to the British in 1833 (Escudé 1986).

The significance of these stories of "territorial loss" is rooted in the post-colonial experience of South America (see Morris 1995). The breakup of the Spanish Empire in Latin America in the first part of the 19th century was the catalyst for the creation of fledgling South American states. Over the course of the last century, states such as Argentina, Chile and Paraguay clashed over the definition of international boundaries and the colonization of the lands of indigenous Indian populations. By the end of the 19th century, most of these nations had introduced formal education systems which included instruction on the history and geography of their particular state (see Escolar et al 1994). Throughout the Southern Cone, teachers were instructed to inform pupils that their national territories had been shaped as a consequence of interaction with aggressive neighbours. For countries such as Argentina and Chile, the colonization of the Antarctic by the British and other extra-hemispheric nations was considered as further evidence for the then widespread perception that these states had been victims of territorial aggrandizement (see Moreno 1960).

The South American examples cited here are emblematic of a recurrent and widespread phenomenon within the SORS as public education continues to promote geographical awareness of the Antarctic and the Southern Ocean. In Australia and New Zealand, particularly in Tasmania and the South Island, polar education is concentrated on the parts of the continent encapsulated by national claims such as the Australian Antarctic Territory. In recent years, the Australian Antarctic Division has been prone to remind citizens of Australia's special relationship to the polar continent: "The ancient physical link between Australia and its southern neighbour is mirrored in the modern Australian Antarctic program, which sees an involvement in the Antarctic as a natural part of our development as a nation" (Australia 1995:4). This description is effective because it effaces the political reasons for Australia's territorial claim to the continent and continued involvement in the Antarctic. Naturalized geopolitical reasoning is not, therefore, the exclusive province of South American states. But as this modest example illustrates, geographical indicators such as proximity and geological continuity are employed to justify and explain polar interests to public audiences in Australia and the southern hemisphere.

These historical and cultural legacies of polar exploration and commerce are usually taken for granted by political leaders and the general public. Whilst many citizens in the southern hemisphere can recall the names of famous explorers and/or expeditions, there is often little critical appreciation of what motivated claims to polar territory or to participation in scientific programmes such as the 1957–8 International Geophysical Year. The significance of this state of affairs is that public education and media-reported stories play a part in the naturalization of polar claims. Claims to a Ross Dependency or a Chilean Antarctic Territory become encoded as a natural constituent of the national geo-body rather than an administrative fiction which most national citizens will never physically see let alone walk over. The minerals-related controversies surrounding the Antarctic in the 1980s powerfully demonstrated the capacity of individual interest groups such as Greenpeace and the Australian Conservation Foundation to mobilize public interest in the polar region and to campaign against government policies. The shift in attitude regarding the future of CRAMRA in 1989 was a dramatic illustration of public opinion influencing a government decision to: "pursue the urgent negotiation of a comprehensive environmental protection convention [and] specifically explore the prospects for the establishment of an Antarctic wilderness park" (Hawke 1989 cited in Darby 1994:116–17). Polar education does not necessarily imply, therefore, that public opinion will uncritically accept government policies affecting the Antarctic and the Southern Ocean.

Middle powers and international relations in the Antarctic region

Contemporary debates in international political economy (IPE) revolve in part around the capacity of so-called "middle powers" to influence the policies and practices of stronger states such as the USA and Japan. Countries such as Australia, New Zealand and South Africa were considered middle powers often on the basis of crude assessments of national power; economic size, population, and/or military strength (e.g. Nossal 1989). Attention focused on these states to determine how trade policies and the international financial system were influenced in the wake of the collapse of the Bretton Woods international economic order and the economic hegemony of the United States. The creation of the Cairns Group of Fair Traders in 1986 was cited as an instrumental case study by these middle powers because it was designed to bring pressure to bear on the Uruguayan round of the GATT talks and in particular on the negotiations for the protection and regulation of their markets within global agricultural trade (Cooper and Higgott 1990/1, Corbridge et al 1995).

The concept of a middle power has received some attention from the disciplines of international relations and political geography (Fox 1980, Holbraad 1984, Nossal 1989). The Canadian scholar, J.W. Holmes, coined the term "middlepowermanship" and noted that Canadian foreign policy was based on "mitigating political tensions and military conflicts Ottawa feared would spark a Third World War" (cited in Cooper and Higgott 1990/1:3). Martin Wight's classic treatise on *Power Politics* asserts that the middle powers of the international system were characterized by their ability to dominate regional issues and agendas (Wight 1995). Paul Kennedy's *Rise and Fall of the Great Powers* also combined historical analysis with a realist understanding of the international system in order to argue that over the last hundred years a number of middle powers have attempted to challenge "pax Britannica" or "pax Americana" (Kennedy 1988), whilst Saul Cohen points to the existence of a number of second-order powers (middle powers) "which are sometimes courted by major powers because of their real or imagined abilities to influence events within or even outside their regions" (Cohen 1984:194, Nierop 1994).

World-systems theorists and neo-Marxist writers on "sub-imperial" powers consider middle powers to be part of the semi-periphery of the world-economy (Marini 1972, Wallerstein 1983, Simon 1991, Taylor 1993). In Wallerstein's spatial representation of the world-economy, the semi-periphery is considered to be of political importance because these nations (Argentina, Australia, South Africa) stabilize the economic extremes of the core (Western Europe, North America and Japan) and the periphery (the Developing World). The categories of core/semi-periphery/periphery were not considered to be static in terms of which nations were included. Under world-systems theory, the semi-periphery is probably the most dynamic politico-economic category (Hopkins and Wallerstein 1982). The SORS would be labelled part of the semi-periphery and, therefore, considered to be important regional components of the world-economy. The semi-periphery states of Latin America and South Africa have been considered as key locales of class struggle, and this has had distinct implications for the political and strategic roles (such as apartheid South Africa during the Cold War) played by these states in the second half of the present century.

This section is not intended to make a major statement on the relative merits of the aforementioned contributions either to realist political economy or to world-system theory. The conceptualization of "middle power" and the international system by scholars such as Wight and Cohen is contestable at least in the sense that the arguments surrounding "middle powers" are usually informed by a realist understanding of the international system which is concerned with the inter-action of states, rational self-interest and the restricted agendas of national interests and security (Buzan 1991, McGrew 1992). On that latter point, it is abundantly clear that power within the international system cannot be adequately understood in the narrow terms of military strength or unilateral action. The development of trans-boundary issues such as migration, refugees and environmental pollution has implications for the agendas and policies of all states. More specifically, the Antarctic raises concerns over global co-operation, governance and power-sharing (Vogler 1995, Anderson 1996).

The arcane language of "middlepowermanship" should not obscure the fact that middle-power diplomacy has been influential whether one considers the international trade negotiations of GATT or the international politics of the Antarctic. For instance, Australia played a key role in the discussions over agricultural fair-trading whilst it was also an influential participant in the Law of the Sea negotiations and the Antarctic minerals discussions in the 1980s. The most effective invocation of middle-power diplomacy in the history of the Antarctic Treaty System occurred during the negotiations over the final version of the Washington Treaty in 1959 when a group of southern hemispheric nations persuaded the United States and Britain to accept a complete ban on nuclear testing in the Antarctic (Scilingo 1973). Prior to this momentous occasion, both these powers had reserved their rights to conduct nuclear tests in the region. The creation of a regional coalition coupled with the combined leadership of Argentina and Australia proved crucial during the Washington negotiations. In that sense "middlepowermanship" triumphed and resolved an issue which threatened to worsen relations between the superpowers given that the Soviet Union had supported a total ban on nuclear testing.

Middle-power diplomacy and regional co-operation between South American and African states was apparent in October 1986 when the United Nations decided to endorse the creation of a South Atlantic Zone of Peace and Co-operation (SAZOPC). The initative for the SAZOPC was due to the endeavours of the Brazilian government of José Sarney. The subsequent Brazilian request to include the proposal for a Zone of Peace and Co-operation on the agenda of the 41st United Nations General Assembly (UNGA) was *inter alia* supported by Angola, Argentina, Nigeria and Uruguay. This resolution was later approved by the UNGA by 128 votes to 1 (the United States). The American government along with a number of European states had complained about the geographical vagueness of the zone of application and the provisions for the management of the region. The decision by the British government to construct Mount Pleasant airbase on the Falklands, coupled with repeated fears that this could lead to the deployment of nuclear weapons, led South American parties to advocate as their top political priority the banning of all nuclear weapons in the South Atlantic (Bartolome 1989, 1991). Despite American objections, the SAZOPC was also linked to the Antarctic Treaty area thus enabling Argentina and other South American allies to apply political pressure to a variety of issues ranging from Britain's decision to hold military exercises (Operation Fire Focus) in the South-West Atlantic to South Africa's illegal occupation of Namibia. Whilst these endeavours have enjoyed moderate diplomatic successes, the underlying strategic rationale for the SAZOPC and the Antarctic Zone of Peace remains the absence of military activity and confrontation from a proximate region (Beck 1988, Albuquerque 1989, Mills 1996).

An interesting dimension of the Brazilian proposal for the SAZOPC in October 1986 was the link made to wider forms of southern hemispheric co-operation:

> There is a wide horizon for co-operation among our countries in the economic, social, technical and scientific fields. There is still a lot to do and the peoples of the South Atlantic with the invaluable aid of the international community and the pertinent multinational institutions, will do whatever is within their reach in order to overcome their own deficiencies and lack of resources, working jointly for the common good. We are resolved to explore the possibilities available to the countries which have self-trust, in their associates and in the future of co-operation between developing countries. (UNGA 1986 A/41/Pv. 50 cited in Bartolome 1989)

A more contemporary example of southern hemispheric co-operation and middle-power diplomacy was the creation of the Group of Temperate Southern Hemispheric Countries on Environment (the Valdivia Group). After the Kyoto Atlantic Treaty Consultative Parties (ATCP) meeting in April 1994, the New Zealand delegation approached a number of states including South Africa, Chile, Argentina and Australia with a view to forging new diplomatic and political connections. It was hoped that these states could promote wider southern hemispheric co-operation in the management of environmental issues. A key sentiment of these earliest meetings in 1994–5 was the belief that recent environmental meetings such as the 1992 Rio Summit had been dominated by the industrial and political interests of northern hemispheric nations (Dalby 1996). Whilst the Antarctic was not on the agenda of the first meeting held in Valdivia in March 1995, the group nevertheless discussed areas of mutual concern such as climatic change, biodiversity, ozone depletion, desertification and forestry. Moreover, given their unhappiness over United Nations environmental groupings, scientific committees were created for the purpose of devising collective recommendations on these areas.

The second meeting of the Valdivia Group was held in Canberra in March 1996. After numerous international meetings, it became apparent that the earlier deliberations on environmental topics had been transformed into an exchange of ideas and policy options when presented at the first Conference of the Parties to the Framework Convention on Climate Change, the Commission on Sustainable Development, Convention on Biological Diversity, the International Whaling Convention, the Basel Convention and the Montreal Protocol. The significance of the Valdivia Group is difficult to assess at this stage given its political infancy, but it does illustrate the determination of middle-power southern hemispheric nations to shape future legislation on global environmental management. As the Australian Minister of Foreign Affairs Alexander Downer concluded in March 1996:

> The Valdivia Group was founded in 1995 to cultivate links across the southern hemisphere, to facilitate the identification of common environmental experiences and interests, and translate such shared interests into joint diplomatic action in the international environmental field. The Valdivia Group is a strategic alliance built around shared objectives. (cited in New Zealand 1996b)

The Valdivia Group is a reflection of an expanding portfolio of regional co-operation. In strategic terms, four members of the Valdivia Group (Argentina, Chile, South Africa and Uruguay) have already committed themselves to upholding the South Atlantic as a zone of peace and co-operation (Venter 1996), whilst the National Unity Government of South Africa has promoted further naval co-operation with its Valdivia Group partners and encouraged neighbouring Namibia to support the SAZOPC. Moreover, South Africa was instrumental in persuading the UNGA to adopt a new resolution in September 1994, calling for the South Atlantic to be recognized as a nuclear free zone and welcoming the entry into force of the Treaty of Tlatelolco for all the states of Latin America and the Caribbean (UNGA 1994 A49/49 cited in Mills 1996).

The capacity of the SORS to represent regional views on the Antarctic and Southern Ocean varies from continent to continent. In the case of South Africa it not only commands the southern African region in terms of conventional indicators but is the only African country with an active interest in the Antarctic and Southern Ocean (Simon 1991, Booth and Vale 1995). It is unlikely that other African states will seek to enter the Antarctic Treaty System in view of pressing social and economic challenges closer to home. Wider African involvement has been limited thus far to the contributions of states such as Kenya, Zimbabwe and Zaire during the United Nations debates in the 1980s on the "Question of Antarctica". In the South Pacific region, Australia and New Zealand are the two most economically and politically important states and the only regional participants in the ATS. Their role within the Antarctic Treaty System has been substantial in terms of shaping the conventions and agreed measures on resource conservation and environmental protection. In terms of active participation in the Antarctic, the most interesting southern hemispheric continent is the South American mainland. During the 1980s, a number of other South American states entered the ATS either as Consultative or Non-Consultative Parties: Brazil, Colombia, Ecuador, Peru and Uruguay. As a consequence of this development, Latin American contributions to the international politics of the Antarctic will increase in significance as future conventions on non-living resource conservation and territorial sovereignty are renegotiated. Alternatively, Argentina's and Chile's capacity to represent the regional interests of South America could fracture as other South American states strive to articulate different agendas.

Geographies of the Antarctic and the southern oceanic rim

Geographers, international lawyers and oceanographers have expounded much intellectual energy attempting to define a Southern Ocean (e.g. Joyner 1990b, Glassner 1993, Broadus and Vartanov 1995). In supposed contrast to the Arctic Ocean, the Southern Ocean is not a semi-enclosed sea and, therefore, has no clearly definable geographical limit. In 1914, the British Admiralty issued guidelines over the geographical delimitation of the Southern Ocean. The northern limit was defined by an imaginary line drawn from the bottom of South America, Africa and Australia, with the southern limit provided by the presence of the Antarctic continent. More recently, the most frequently cited northern limit is the Antarctic Convergence (also called the polar frontal zone), an area situated between 55 and 52 degrees South latitude, where the polar waters converge with the warmer waters of the Atlantic, Indian and Pacific Oceans (Glassner 1993:98). Other writers have argued that the Southern Ocean terminates at 40 degrees South latitude because the Antarctic covergence was thought to demarcate a sinking of the northward-moving Antarctic surface water below the southern-moving sub-Antarctic water (eg Broadus and Vartanov 1995). Whatever the merits of these definitions, the only geographical definition that has any substantial international legal status is 60 degrees South latitude, the formal jurisdiction area of the Antarctic Treaty System. Since 1961, the Antarctic continent and the surrounding seas have been governed by this international regime which, amongst other things, has promoted scientific co-operation and multilateral resource management.

For the purpose of this analysis, the southern oceanic rim is understood as the area encompassing the Antarctic Treaty region and the SORS that reside proximate to that region (see Map 1). Proximity is stretched to include states over 3000 miles from the Antarctic continent and hence the concern for the activities and polar perspectives of states such as Australia, India and South Africa. This short review is intended to set the scene for the more detailed examinations of individual nation-states and their involvement in the Antarctic and the Southern Ocean. Three major themes have been identified which have had, and continue to have, profound implications for the geopolitics of the Antarctic: the Antarctic Treaty System, territorial sovereignty and the Law of the Sea.

Legal geographies of the Antarctic Treaty System

In 1961, the Antarctic Treaty System ratified a legal and political regime which has transformed the international politics of the Antarctic below 60 degrees South latitude. The terms of the Antarctic Treaty define the polar region as:

> The provisions of the present Treaty shall apply south of 60 degrees South Latitude, including all ice shelves, but nothing in the present Treaty shall prejudice or in any way affect the rights, or the exercise of rights, of any State under international law with regard to the high seas within that area.

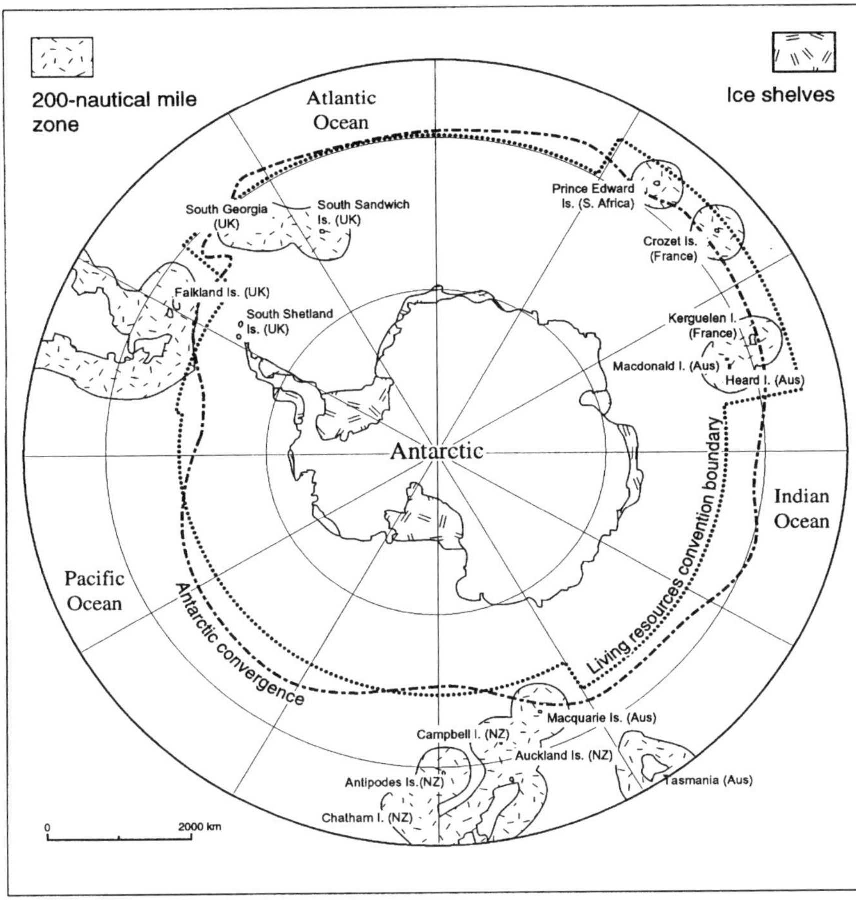

Map 1 Zones of jurisdiction in the Antarctic and Southern Ocean. Reprinted from *Applied Geography* **10**, Maritime Zones in the Southern Ocean, p. 308, with kind permission from Elsevier Science Ltd., The Boulevard, Langford Lane, Kidlington, Oxford, OX5 1GB, UK

There have been misgivings that this boundary is an arbitrary one and hence lacks a biogeographical basis (Auburn 1982:129, Joyner 1990b:314). International lawyers have argued that the zone of application not only bisects the Southern Ocean but also excludes large areas of the Antarctic convergence. If the Antarctic Treaty had encapsulated a larger maritime area, a greater geographical area would have been under its mangement. Most objections to the area of application in the Antarctic Treaty are based on a rather limited understanding of geography: since to assume that there are natural boundaries for the Southern Ocean is to ignore the inference that our designations and categories are natural and hence uninfluenced by social and political criteria. Regional boundaries, whether for the Southern

Ocean or the Antarctic continent, are deliberate inventions imposed upon the earth's surface. More valuable, therefore, is the significance attached to these particular designations rather than the merits of geographically correct boundaries. The failure of the 1959 Antarctic Treaty to define the precise geographical co-ordinates of the Antarctic continent was rooted in fears pertaining to the legal and political status of ice shelves, rights to the high seas and the importance of floating ice (Beck 1986, Watts 1992). It was not until the advent of remote sensing and satellite technology in the 1970s that the first complete image of the Antarctic and its changing ice shelves was produced for the human gaze (Dodds 1997).

The most significant provisions of the Antarctic Treaty for the SORS relate to the banning of military activities, the freezing of sovereignty and the complete prohibition of nuclear testing. The creation of an inspection system under auspices of the Antarctic Treaty, whilst initially resisted by Argentina and Chile, instilled confidence as every party could henceforth appoint observers to carry out on-site inspections of others' bases and research facilities including boats and aircraft. The provisions of Article IV of the Antarctic Treaty were central to the claimant states such as Argentina and Australia because they ensured that territorial claims were held in abeyance, ie existing claims were protected whilst no claims could be made to polar territory, territorial seas, contiguous zones and exclusive economic zones. The Antarctic Treaty System has guaranteed that the international politics of the Antarctic have, in the main, been dominated by scientific endeavour and environmental conservation rather than national rivalries and unregulated resource exploitation (Beck 1986, Peterson 1988, Chaturvedi 1996, Stokke and Vidas 1996).

The Consultative Parties of the ATS, including the 12 original members of the Treaty, meet annually (previously it was biennially) in order to discuss recommendations for the Antarctic continent or the implementation of existing agreements. Over the past 35 years, scores of recommendations and measures have been created and accepted by the ATS (Joyner 1985:2). These include the 1964 Agreed Measures on the Conservation of Antarctic Fauna and Flora, the 1972 Antarctic Seal Convention, which limited commercial exploitation of that species; and the 1980 Antarctic Marine Living Resources Convention (CCAMLR, enforced in 1982) designed to encourage marine conservation especially of krill stocks in the Southern Ocean. The Convention on the Regulation of Antarctic Mineral Resource Activities (CRAMRA), accepted in June 1988, was intended to provide an institutional framework for the regulation of exploration and exploitation of land or offshore-based activities. CRAMRA was later rejected in 1989 when two claimant states, Australia and France, raised fundamental objections to the proposed minerals regime. More recently, the ATS parties have successfully negotiated a Protocol on Environmental Protection to the Antarctic Treaty (Protocol) to replace earlier attempts to negotiate a minerals-based regime. The major difference between these two documents is that the Protocol makes the environmental protection of the Antarctic the highest priority of the ATS and prohibits all mining and mineral exploration (which CRAMRA intended to permit with strict reservations) in the region (Wolfrum 1995).

The geographical jurisdiction of the Antarctic Treaty System has been extended as a consequence of specific recommendations by Consultative Parties. Whilst the Convention for the Conservation of Antarctic Seals uses the 60 degrees South line, others such as the Convention on the Conservation of Antarctic Marine Living Resources (CCAMLR) employ a line roughly approximate to the Antarctic polar front, thereby including in its provisions for the regulation of krill and fish exploitation some areas north of the 60 degrees line. The International Whaling Commission (IWC), charged with enforcing the International Convention for the Regulation of Whaling (independent of the ATS), established an international whale sanctuary in 1994 for the region covered by the CCAMLR and some geographical regions north of 40 degrees close to the South American mainland (Vogler 1995). Under United Nations Resolution A/RES/41/11, the South Atlantic Zone of Peace and Co-operation was defined using 60 degrees South as the southern boundary of the geographical area of application (Carasales 1989). Despite all this, the 60 degrees South boundary remains the most important under existing international polar law. North of the ATS region, states possessing recognized sovereignty over sub-Antarctic islands within the CCAMLR region (eg South Africa and the Prince Edward Islands) can substitute national regulation and national enforcement within a 200-mile Exclusive Economic Zone if, and only if, these stipulations are equivalent to those found within the CCAMLR (Joyner 1990a:44).

Old geographies: territorial divisions

The Antarctic continent is a paradoxical place for the political geographer. Parts of the continent have been subjected to numerous claims whilst other sections remain totally unclaimed (Map 2). Between 1900 and 1943, seven countries including Argentina, Australia, Chile and New Zealand pressed claims to parts of the Antarctic. States such as India and South Africa, the United States and the Soviet Union do not recognize these claims but accept that the evolution of the ATS since 1961 has established a mechanism for regulating the rights of claimant and non-claimant states.

The so-called "sovereignty problem" in the Antarctic has been one of the most enduring political and academic dilemmas (Peterson 1980, Luard 1984). There were, for example, unpleasant incidents between Britain and Argentina in the 1950s over the issue of territorial ownership (Beck 1989, Dodds 1994a). Before the 1959 Antarctic Treaty, growing fears that superpower rivalry might lead to the militarization of the polar continent

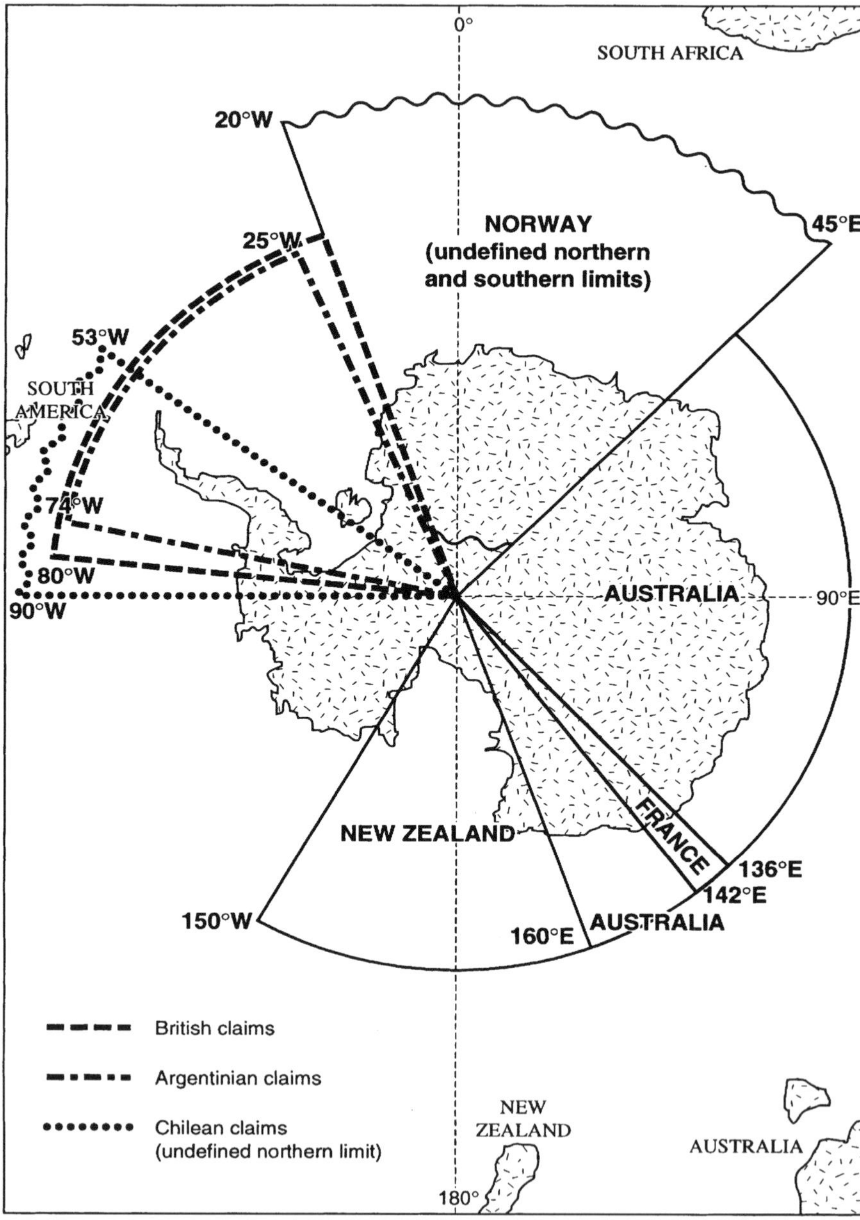

Map 2 National claims in the Antarctic

abounded. Sovereign claims to territory undoubtedly prevented early attempts by the American government in 1948 to institute a new international condominium over the Antarctic. Resistance from SORS such as Argentina, Australia and Chile was a crucial stumbling point (Grolin 1987:40–2). Other proposals for changing the international political regime for the Antarctic (under UN authority) put forward by India and New Zealand in the 1950s were similarly rejected by the claimant states. In each case, the claimant states were not prepared to relinquish their sovereign rights, and the non-claimant states such as the United States were not willing to suspend their rights to make future claims to the polar continent.

The turning point for the management of competing sovereign claims was the 1957–8 International Geophysical Year (IGY). It was agreed, with substantial misgivings from *inter alia* Argentina and Australia, that participating nations could locate their research stations anywhere on the Antarctic continent regardless of prior claims to national territory (Beck 1986, Hall 1986). Five non-claimant states participated in the research programmes of the IGY: Belgium, Japan, South Africa, the Soviet Union and the USA. The Soviets created a number of bases inside the Australian Antarctic Territory and the South Africans took over a base in the Norwegian sector. More than 40 research stations and observatories were created during this period and thousands of scientists were located on the continent (see Beltramino 1993). This agreement did not, however, undermine the central importance of the sovereignty problem; rather it acted as an illustration that international scientific co-operation was possible in the face of these concerns. It should be remembered that the IGY polar programme was only possible because an agreement reached in a preparatory meeting in Paris in 1955 ensured that sovereign rights in the Antarctic would not be affected by these scientific programmes (Hall 1986).

The IGY, whilst of great scientific value, was not designed to resolve the issue of competing sovereignties in the Antarctic and the Southern Ocean. Seven states retained their claims to the Antarctic continent throughout the IGY and only New Zealand ever offered to drop its polar claim. The other five IGY participants refused to acknowledge the validity of these polar claims. The 1959 Antarctic Treaty did not resolve this matter either: it simply suspended the whole question. The legal status quo was effectively frozen (Beck 1986, Peterson 1988). Claimant states were not required to give up their claims, but no new claims or the enlargement of existing ones would be allowed. The Treaty also stipulated that no activities taking place whilst the treaty was in force could form the basis for asserting, supporting or denying any territorial rights on the Antarctic and surrounding oceans (Grolin 1987:42). The most desirable solution might have been a complete relinquishment of all territorial claims in the region, but as the New Zealand delegation to the 1959 Washington Conference noted:

> In theory the ideal basis upon which to formulate a treaty would be for former nations to abandon their claims and for the latter nations to renounce the right they reserve to make a claim. This solution, however, would envisage the establishment of a sovereign authority to administer Antarctica and would require a degree of national disinterestedness which the Washington talks have shown just does not exist. (cited in Dodds 1996c)

The subsequent legal and political flexibility of the Antarctic Treaty has been of sufficient quality that long-standing territorial rivals such as Argentina and Britain have been able to co-operate in polar matters in spite of later events such as the 1982 Falklands/Malvinas War. Many commentators agree that the issue of sovereign rights in the Antarctic remains important, and claimant states such as Argentina, Chile and Australia have undoubtedly carried out acts of administration (or bizarre stunts such as flying pregnant women to Antarctic bases) designed to register their sovereign interests in the polar continent. Sovereign claims to polar territory have also frequently prevented stronger recommendations being enforced for environmental and or resource protection in the region. The Convention on the Conservation of Antarctic Marine Living Resources was intent on establishing resource management for the Southern Ocean in the face of fears of over-fishing by nations such as Japan and or the Soviet Union (Mitchell and Tinker 1980). Whilst many ecological commentators praised the intent of the CCAMLR, there was criticism of the enforcement mechanism for stock conservation and fishing levels. The regulation of fishing around South Georgia has been a striking illustration of the problems posed by territorial sovereignty. Conservation measures have been blocked by fishing nations such as Russia and Japan, whilst Britain and Argentina dispute each other's rights to issue licences for fishing in the first place. A subsequent proposal by the United States to create a fishing commission to enforce fishing catches and to carry out inspections of trawlers was rejected by Argentina and Chile who argued that their maritime and polar sovereignties would be compromised (Orrego Vicuna 1990). Ultimately, as Grolin concluded "a so-called 'bifocal approach' was adopted whereby the claimant and non-claimant states could interpret the reference to 'coastal state jurisdiction' in Article IV each in their own way" (1987:46). This kind of compromise reflects long-standing difficulties over competing or co-existing sovereignties inside the ATS.

The critical theme following the minerals negotiations of the 1980s centred on the accommodation of territorial sovereignty within any new legal regime. Rolf Trolle Andersen, the Norwegian negotiator, neatly expressed the dilemmas facing the various diplomatic parties:

> It was apparent from an early stage that a mere repetition of the provisions of the Article IV of the Antarctic Treaty would not be sufficient. The ownership of the sub-soil is so directly connected with the concept of territorial sovereignty that the claimant countries needed a more concrete manifestation of their

> position. At the same time any granting of a preferential position for the claimant countries by a direct reference to their claims could be interpreted as prejudicial to the non-claimant views. What the parties thus had to do could be described as squaring the political circle. (cited in Kyriak 1996:10)

The difficulties facing the negotiators were further compounded by issues of geographical location, developmental profile, political allegiances, economic interests, and existing attitudes towards the wider interests of the international community. CRAMRA had, for example, intended to secure the consent of the claimant states on the basis of articulating special positions in the negotiating process whilst simultaneously protecting the interests of important non-claimant states such as the United States. The legal and political environment for CRAMRA was far removed from the earliest negotiations over the Antarctic Treaty. In 1959, when few nations appreciated the geopolitical and resource value of the Antarctic, there was only a limited interest from the newly emerging countries of Africa and Asia. By the early 1980s, this position had altered and territorial sovereignty in the Antarctic had become a far broader international concern. The role of the United Nations during the 1980s, for instance, was indicative of this development as the member states of the ATS were forced to account for their negotiations and to defend themselves against the claim that the Antarctic should be governed by the UN General Assembly rather than the ATS. The growing profile of the environmental movement in the 1980s also raised profound questions over who had the right to determine the fate of the Antarctic continent (Darby 1994). The role of Greenpeace and the Antarctic and Southern Ocean Coalition (ASOC) was crucial during the minerals negotiations and, once again, the political legitimacy of the claimant states and the ATS more generally was being questioned by the wider international and ecological community.

At the heart of recent controversies such as the minerals negotiations were territorial sovereignty and resource ownership. The chairman of the negotiations, Chris Beeby, had argued in October 1988 that the final Convention on minerals was historic because:

> All too often the international community has come to grips with major problems at a very late stage in the game. It tends to react after catastrophes have occurred, wars have been fought or the environment has been seriously degraded by human activity. The Antarctic Treaty Parties have, however, had the good sense to put binding rules in place well in advance of any activity and the political and environmental disruption, which would, almost certainly, have resulted in the absence of such rules. (cited in Kyriak 1996:70)

The Convenor of ASOC in New Zealand, Cath Wallace, argued that the Convention on minerals was flawed because:

> By providing property rights the Convention will facilitate mining and the introduction of rivalry over strategic resources in the zone. It will introduce new and dangerous stresses to the Treaty and area. This would endanger not only the

> environment but could lead to conflict and destabilise the Treaty and the zone's demilitarised status . . . Diplomats and environmentalists because of their different standards of reference, differ in their assessments of the efficacy of the environmental protection afforded by the proposed minerals convention. (cited in Kyriak 1996:71)

Whilst the environmental opposition movement against CRAMRA proved highly effective in mobilising opposition to its provisions, the subsequent Protocol did not resolve any of the issues regarding territorial sovereignty. At the heart of Wallace's objections to CRAMRA was the issue of ownership and whether the Antarctic should be considered a place under the sole jurisdiction of a select group of states. The ensuing negotiations of the Protocol and the introduction of procedures and mechanisms designed to protect the polar environment will depend, once more, on the ability of states to accept effective enforcement procedures. An external organization such as the CEP, responsible for application and enforcement, will inevitably be regarded with suspicion by claimant states such as Argentina and Australia because it would be prejudicial to their sovereign authority. The prospects for a fundamental change in attitude appear bleak, as southern hemispheric nations in particular invest considerable monies in supporting their claims.

New geographies: Antarctica and the Law of the Sea

The evolution and application of the Law of the Sea (LOS) to the Southern Ocean has been fraught with difficulty, as the provisions of the Antarctic Treaty clash with some of the fundamental principles of the LOS. The Antarctic Treaty was negotiated in the immediate aftermath of the first LOS conference in 1958 (Couper 1978, Glassner 1990). In that sense, the Antarctic came of age politically and scientifically along with the ocean floors and outer space. The United Nations Convention on the Law of the Sea (UNCLOS I) established legal principles concerning the lawful rights and duties relating to the high seas, the territorial sea, fishing and the management of resources and the continental shelf (Glassner 1990). The physical geography of the oceans and the seas was divided into discrete categories which were each invested with political and legal importance. Under the UN Territorial Seas Convention coastal states can claim a territorial sea up to 12 nautical miles from their coastline, which can then form the basis for resource exploitation, and, a 24-mile contiguous zone which could involve a relevant state in the regulation of custom and or immigration regulations (Joyner 1990a:318).

At that stage in the late 1950s, the application of the provisions of the LOS was largely concentrated on the immediate maritime areas of sovereign states rather than the ocean floors, the Antarctic or outer space. Throughout the 1960s and 1970s, most of the international community had minimal interest in these regions. The LOS negotiations of the 1970s were dominated by discussions over the definition of the continental shelf and the rights of maritime states to exploit natural resources. Whilst the Sri Lankan delegate to the 1975 LOS negotiations raised the issue of bringing the Antarctic to the attention of the delegates, most of the meetings were dominated by those aforementioned issues rather than the Antarctic or deep seabed (Joyner 1992). Growing interest in the future of the Antarctic coincided with the most important conference on the LOS, concluded in Montego Bay, Jamaica in December 1982. The implications generated by this body of international law regarding the ocean floors and continental shelves for the Antarctic and surrounding oceans were substantial. The 1982 United Nations Convention on the Law of the Sea (UNCLOS III) added, substantially, to the existing canon of international maritime law (Triggs 1986, Joyner 1992). The provisions of UNCLOS III had a direct bearing on the Antarctic Treaty System as coastal states were invested with new powers to declare 200-mile economic exclusive zones (EEZs) whilst the resources of the deep seabed were a "common heritage of mankind" ie beyond the jurisdiction of individual states. The creation of 200-mile EEZs' enabled states under the provisions of UNCLOS III to declare sovereign rights over natural resource activities within an EEZ. The 1982 Convention also declared that the continental shelf was now considered to extend to 200 miles off the coastline regardless of whether this was physically the case. In doing so, it dealt with the thorny issue of coastal state rights over the area of sea bed which lay within an EEZ but was beyond the edge of the continental shelf. The 1982 deliberations effectively reversed the original findings of the 1958 UNCLOS proceedings and also extended the rights of land-locked states to resources under the legal and political rubric of common heritage (Herber 1991, Joyner 1992, Glassner 1993).

The evolving body of the Law of the Sea had undoubtedly complicated the legal and political geographies of the Antarctic and Southern Ocean (Joyner 1992). Whilst the USA has refused to ratify the provisions of UNCLOS III, it has recognized the implications for the Southern Ocean. The establishment of an exclusive economic zone depends on the adjacent territory being controlled by a recognized sovereign coastal state; and in the case of the Antarctic, the seven would-be claimants are not only unrecognized by other states but are also effectively suspended under the provisions of the 1959 Antarctic Treaty. Moreover, the polar claims of Argentina, Britain and Chile overlap and are therefore subject to competing sovereignties. Antarctica is, thus, neither a continent composed of sovereign coastal states nor a recognized condominium possessing a uniform coastline frontier (Joyner 1990b:320). The few attempts to assert an exclusive economic zone (eg Australia over the AAT in 1979) have been reversed as a result of adverse international reaction. The widespread non-recognition of polar claims means that it is a contestable point as to whether claimant states could assert authority over adjacent waters and continental shelves. At present, the provisions of the Antarctic Treaty have effectively suspended (but not eliminated) declarations of sovereign maritime jurisdictions.

UNCLOS III's decision to label the resources of the deep seabed a common heritage of mankind had implications for the southern polar region. The seabed is effectively considered to be beyond the national jurisdiction, including rights of appropriation, of any one state. The provision for the creation of an International Seabed Authority to supervise the resource exploitation of the deep seabed was controversial given that the revenue derived from the resources would be distributed amongst the international community regardless of which nations carried out the initial prospecting and exploration, and which would supervise resource exploitation on the deep seabed. A number of states, notably Malaysia, were quick to link the provision between the deep seabed and the Antarctic. Hence the concern of the Chilean delegate Fernando Zegers in August 1980 that the control of Antarctic resources could be jeopardized by the UNCLOS negotiations. He subsequently argued in a secret meeting of southern hemispheric nations that:

> The superpowers have a very strong interest in maintaining the Antarctic Treaty. Their only alternative to an Antarctic Treaty solution is a common heritage solution. Provided the economic concessions to claimants are approximately of the same order as would otherwise be forfeited to the common heritage then political interests will continue to motivate them to see an Antarctic Treaty solution. (New Zealand archives, Prime Minister's Department 1980)

Malaysia and Sri Lanka have argued that the Antarctic should be declared a common heritage of mankind on the basis of the precedent set by UNCLOS III. Antarctica should be: "made subject to a regime of rational management and utilisation to secure optimum benefits for mankind as a whole and, in particular, for the developing countries, in accordance with appropriate global international arrangements, and within the frame-work of the new international order" (Sri Lanka 1981 cited in Triggs 1986:277). If the principle of common heritage was ever universally accepted then Antarctica could be legally transformed into a *terra communis* and existing national claims would be considered redundant (Triggs 1986:278). The most powerful legal argument in favour of common heritage is the principle of customary law as has already applied to other uninhabited geographical areas such as outer space and the moon (Herber 1991, Vogler 1995).

An ironic feature of these earliest debates about common heritage during 1983–4 not widely recognized was the role India played in supporting Malaysia's proposals to the UN General Assembly. Having been invited to join the ATS in 1983, it was anticipated that India would not press the case for common heritage. Instead it is believed that the Indian government encouraged Malaysia to assess the extent of international support for a declaration of Antarctica as a common heritage site. The New Zealand government was later instrumental in persuading India to join the ATS and to refrain from publicly promoting common heritage.

The UNCLOS III convention also:

> stipulated that islands are legally capable of generating their own 12 mile territorial sea, 200 mile EEZ, and extensive continental shelf delimitation . . . In the sub-Antarctic part of the Southern Ocean, there are certain undisputed mid-ocean island groups whose legal status is relevant to circumpolar maritime zones . . . Their presumed worth lies in the sphere of scientific research upon the Antarctic Convergence . . . but the islands' real value derives from their role as legal hubs for the extension of marine resource jurisdiction into the Southern Ocean. (Joyner 1990b:321)

The two mid-ocean island groups, Prince Edward (South Africa) and Heard and McDonald (Australia), are indicative of the problems of extended maritime jursidictions. In the midst of the UNCLOS negotiations, the Australian and South African governments declared 200-mile fishing zones around these island groups. Whilst the Australian government exempted the waters surrounding the Australian Antarctic Territory in 1979, the LOS was used in 1995 by the Keating government to lay claim to the marine resources within the EEZ of the Australian Antarctic Territory.

The legal and political arguments surrounding the Law of the Sea negotiations have placed the SORS in an awkward position. Although Argentina and Chile accepted the general merits of common heritage and the New International Economic Order (NIEO) they have been ardent critics when these same principles were applied to the Antarctic (Parkinson 1985, South Commission 1987). As the Chilean delegation to the United Nations in 1984 noted: "To tamper with the system would be to endanger all that has been achieved and to return to a bygone age; this would have a negative impact on mankind as a whole . . . the Antarctic system, its authority and practical functioning [should be] preserved intact" (cited in Parkinson 1985:450).

Summary

The Southern Ocean is unique. There is little consensus on its exact geographical boundaries as the very zones established by the ATS for the management of resources and peaceful activity are challenged by the common heritage critics of the international community. Unlike any other ocean, a large proportion of the Southern Ocean has been removed from military competition and nuclear testing with the support of the international community and the superpowers. The recent controversies over French nuclear testing in the South Pacific have revealed the fragility of other regional peace zones. National maritime and territorial jurisdiction has also been curtailed under the provisions of Article IV of the Antarctic Treaty. SORS such as Argentina and Australia are prevented from claiming exclusive economic zones within Antartida Argentina or Australian Antarctic Territory. The economic potential of the Southern Ocean and the Antarctic is well appreciated but thus far international agreements have prevented *inter alia* the exploitation of offshore oil and gas (Cook 1990). The environmental security of the Southern Ocean is, therefore, closely related to the environmental provisions affecting the Antarctic continent (Broadus and Vartanov 1995:190). Competing national sovereignties have often been instrumental in "filtering down" the more robust recommendations relating to environmental protection, over-fishing and mineral exploration. A recent example of this would be the provisions of the 1991 Madrid Protocol relating to marine pollution which exempts the activities of naval vessels from strict environmental regulation.

The interests and activities of the SORS in the Antarctic and Southern Ocean revolve around the varied historical legacies of polar exploration, the political geographies of the region, geographical proximity and the significance of public education and the mass media. Throughout the forthcoming chapters, these varied dimensions will be investigated and evaluated.

2

Critical geopolitics and the geo-graphing of Antarctica

The French geographer Yves Lacoste reasoned that: "in international relations, what we call 'Geography' is subjected to two contradictory appraisals. The influence of factors considered to be geographic is either greatly exaggerated or nearly overlooked, in spite of the obvious territorial imperatives" (Lacoste 1984:213). The emerging field of critical geopolitics has argued that the spatialities of international relations constitute an important if neglected research field (Dalby and Ó Tuathail 1996). In contrast to earlier geopolitical writings which tended to treat space as an unchanging backdrop to international affairs, critical geopolitical writers have reasoned that international relations are constituted in place-specific contexts and that the construction of place characteristics is an important feature of practices such as foreign and security policy (Lacoste 1976, Ó Tuathail 1992, Agnew and Corbridge 1995). Critical geopolitics aims "to challenge the assumptions and conceptual infrastructure that have historically governed thought on geography and international relations" (Ó Tuathail and Dalby 1994: 513). In the process, it has become apparent that the aims of critical geopolitics are related to other areas of contemporary research concerned with the production and circulation of geographical knowledge (Godlewska and Smith 1994, Hooson 1994, Agnew and Corbridge 1995, Radcliffe 1996, Radcliffe and Westwood 1996).

The connection between critical geopolitics and the histories of geographical knowledge forms the backbone to this chapter. It has been widely recognized that geography as a field of knowledge is intimately linked to the exercise of power. Edward Said has argued, for example, that imperialism is: "an act of geographical violence through which virtually every space in the world is explored, charted, and finally brought under control" (1993: 225). Geographers, map-makers and surveyors played an instrumental and well understood role in the collection and dissemination of practical geographical information. In virtually every place the Europeans and their allies conquered and colonized, territorial claims were pressed and legitimatized through topographic surveys, evaluations of natural resources, reports on indigenous peoples and the charting of coastlines. The Antarctic continent and the Southern Ocean were not exceptional in this global geographical process. The gradual incorporation of the Antarctic into the European geographical imagination was enabled by those acts of geographical colonization. Changing technologies of exploration and communication such as aerial survey later furthered the comprehensive mapping of the Antarctic and surrounding oceans.

This chapter, therefore, considers the changing location of the Antarctic within the modern geographical imagination. The first section briefly examines the emerging field of critical geopolitics and the changing significance of place. The specific purpose of this discussion is to explain how my account has been influenced by critical geopolitical writings (see O'Loughlin 1994). The second and more substantial section focuses on the changing geographical representations of the Antarctic and the Southern Ocean. A number of key moments in the present century have been identified, illustrating how political understanding of the Antarctic has changed as geographical knowledge altered and new political and economic issues emerged. The central argument of this section is that so-called "naturalized" geopolitical reasoning played an increasingly important role in the political discourses of southern hemispheric governments and media (Agnew and Corbridge 1995). From the 1940s onwards, the Antarctic region was represented within the southern claimant states as a natural extension of their national territories. This kind of geographical argument was particularly prevalent in Argentina during the Peronist era (Child 1988, Dodds 1993). In Argentina and Chile geopolitics became increasingly considered to be a form of hard-headed materialism intent on exposing the political realities generated by the permanent realities of the earth (Hepple 1992, Ó Tuathail 1996). Public education further contributed to the reproduction of territorial claims and resource rights through the medium of geography and history lessons (Escudé 1992). The final part of this chapter draws together these themes and highlights the importance of the Antarctic within an increasingly globalized world.

Critical geopolitics

The term "critical geopolitics" was introduced in the mid-1980s to signify a movement within political geography dedicated to the rethinking and reformation of geopolitics (Ó Tuathail 1996). This appeared highly appropriate given the observations of Geoffrey Parker in 1985 that:

> the adjective 'geopolitical' has, in recent years, assumed considerable popularity with politicians, academics and journalists alike. Almost overnight, as it were, everything appeared to gain a 'geopolitical' dimension, this being understood in general terms to signify an awareness of the importance of geographical factors in human affairs. Mention of the term nowadays is calculated to conjure up visions of oil supplies, strategic minerals, agricultural potential, dangerous sea routes, vulnerable frontiers and, possibly, dwindling natural resources. It is no bad thing that the importance of geography has thus come to be implicitly recognised so widely, but more precision on the exact subject of the study becomes even more important. (Parker 1985:2)

However, in contrast to these recent appropriations of the term "geopolitical", critical geopolitics aimed to interrogate not only the production and significance of geopolitical discourse but also to suggest alternative ways of considering international political space (Dalby and Ó Tuathail 1996).

Critical geopolitics aims to "unfold space for thought" on power, geography and international politics in the face of growing intellectual and political disenchantment with state-centric reasoning, the naturalization of geography and so-called "orthodox" geopolitical reasoning (Ó Tuathail and Dalby 1994). This view seems particularly pertinent in the face of rising political and economic globalization and the growing challenges to the modern sovereign state (Smith 1994, Campbell 1995). For some scholars, such as Michael Mann and Michael Shapiro, geopolitics is emblematic of a geopolitical imagination organized around the modern state system (Mann 1993, Shapiro 1994). Geopolitics has also been considered intimately bound up with the nation-state and its capacity to produce, regulate and survey political space (Agnew and Corbridge 1995, Luke 1996). Moreover, the academic history of geopolitics is usually considered as a catalogue of writings on state power stretching from Mackinder to the present day. In contrast, critical geopolitics has resisted the assumption that the "geopolitical" should be understood as either exclusively bound up with the fixed spaces of the international system or as the exclusive province of a select number of disciplinary authors. One of the most striking aspects of recent episodes of financial globalization, for example, has been the demonstrable capacity of financial flows and fast-moving capital to undermine traditional claims to sovereign power and territorial control (O'Brien 1992, Agnew and Corbridge 1995, Corbridge, Martin and Thrift 1995).

Until quite recently, it was still common for political geographers to produce geopolitical analyses of conflict on the basis of a limited understanding of the role of geography. The latter would have been interpreted as a: "set of external facts entering separately into the origins and course of each conflict or incident . . . What would be missing, however, would be an understanding of the nature and origins of the geographical attributions that are implicit in the reasoning of those who brought about the conflicts and incidents" (Agnew and Corbridge 1995:46). In the case of the 1982 Falklands War, most of the Argentine and British geopolitical analyses identified notions such as "geographical proximity" as being important in Argentina's decision to invade the Islands without ever problematizing the seemingly common-sense understandings of location, distance and proximity (see Hepple 1988). This is now changing as scholars attend to the processes whereby places are depicted within the practices of foreign policy rather than assuming those places to be "exhibits" that merely await the gaze of the geopolitical observer. To put it another way, political geographers are acutely aware that place is not simply an "unchanging or permanent factor" in the practices of foreign policy. Space, as John Agnew (1993) reminds us, has been interpreted by many political geographers as referring to rather fixed understandings of the term. Space is identified as territorial and usually understood as a series of blocks defined by national territorial boundaries. Alternatively, space has been understood as structural whereby geographical entities are seen to have particular spatial effects that result from their inter-relationships with one another. What is often lacking in these formulations is an understanding of the historical and political circumstances that shaped particular understandings of place (see Sidaway 1994, Smith 1994, Agnew and Corbridge 1995).

In contrast to the implict assumptions of conventional geopolitics, space is not considered either as absolute or simply a backdrop to the unfolding of international politics. At this point, two important elements of the critical geopolitical project deserve some elaboration. The first relates to a critique of forms of epistemic realism and the second to political realism. Following Michel Foucault's warning that the world does not present itself with a "legible face", critical geopolitical writings have focused on how varied representations and interpretations of the world influence and affect subsequent judgement and action (Foucault 1984). Contrary to the claims of epistemic realism, critical geopolitical writers argue that our representations of the world, and our understandings of the Antarctic for that matter, are neither self-evident nor do they exist in a world independent of ideas and beliefs. Geopolitical discourse refers to the modes of representation which structure assumptions and to understandings and explanations about the political world. The term "geo-graph" was introduced into the critical geopolitical domain by the work of Gearoid Ó Tuathail and Simon Dalby. In his earliest writings, Ó Tuathail argued that the term signified the capacity of geography to produce particular representations of the earth's surface through the technologies of mapping and classification (Ó Tuathail 1989, 1996). Moreover, the subsequent grids of intelligibility imposed upon those representations of the earth's surface play an important part in the construction and reproduction of geographical space.

The key feature of these critical geopolitical investigations of the geographing of place was to insist that places were not simply waiting to be interpreted by the disinterested and distant geopolitical observer. In conjunction with recent writings on histories of geographical knowledge, critical geopolitics has focused on the various institutions, practices and sites which have rendered places "visible". This criticism has encompassed the discursive practices used to delimit global political space, the institutions such as government ministries and the media which enable the interpretation of places and, finally, the literatures and political conventions (such as Sovietology or Development) which enable places to be evaluated and judged (Dalby 1990, Escobar 1995, Ó Tuathail 1996). During the Cold War, numerous Western-based institutions and individuals produced wide-ranging geopolitical surveys of global political space including Antarctica and the Southern Ocean. In contrast to the often-stated claims that these surveys were merely recordings of manifest realities, critical geopolitical writers have deconstructed those claims in order to investigate the hegemonic ways of seeing, siting and citing place (Dodds and Sidaway 1994).

The subsequent discussion of the geo-graphing of Antarctica is based on the assumption that changing understandings of place have an important impact on the international politics of the region. This account is also sympathetic to the earliest critical geopolitical writings on the construction of place within political discourse. As Dalby and Ó Tuathail remind us: "Places constituted in political discourse need not be stable to be politically useful; multiple narratives can sometimes render a particular place or state in a number of ways simultaneously . . . These devices need not be very sophisticated to function in political discourse . . . repetition of themes is an effective method of constructing identity" (Dalby and Ó Tuathail 1996: 452). The periodization of the Antarctic is not meant to imply that the Antarctic was understood in radically different ways as new issues such as resource management materialized but that the continuities and discontinuities are equally important in contributions to political understandings of the Antarctic and the Southern Ocean.

Political realism refers to a corpus of ideas and texts that have formed an important set of assumptions about the international political arena. Whilst there are significant differences within political realism, there are also some core themes that have a direct bearing on geopolitics. International political space is considered to be composed of territorially bounded political communities co-existing in an anarchical world. International politics is considered analytically and politically distinct from domestic politics. As Anthony McGrew has usefully noted:

> Traditional geopolitical thinking constructs the globe as a single strategic space in which states compete for power and advantage. With no world government or authority to protect them, states are led to acquire power in order to protect themselves against potential threats. Security and power are thus intimately linked . . . Security is [also] conceived primarily in territorial terms – that is, ensuring the territorial integrity . . . In this geopolitical construction of the world, states are conceived almost exclusively as territorially bounded power

> containers. As such . . . Traditional geopolitics is concerned with the relationship between space, place, territorial security and state power. (McGrew 1992:17 but also Klein 1994)

Further, the nation-state is considered to be the most important political actor, and the pursuit of national security is interpreted as a priority for individual states (Booth 1991, Dalby 1994). Security in this context has usually been thought of as referring to military security and the importance of protecting a domestic population from external dangers (Campbell 1992). Recent debates over environmental security have illustrated a considerable capacity for broadening debates *inter alia* about environmental protection and sustainable development. In relation to the last point, political actors such as social movements and global organizations have not been accorded enough importance in realist accounts of international politics. The international politics of the region during the 1980s, for example, were shaped by the activities of ecological movements and the United Nations.

Critical geopolitics has explicitly rejected realist accounts of international politics because it is a partial and incomplete account of global political life. There are other forms of social and political activity which cannot be accounted for or colonized by the nation-state. As Paul Routledge has argued:

> Attention to the actions of social movements would contribute to a critical geopolitics in at least two ways. First, it would (de)centre analytical focus away from an exclusive concern with the machinations of the state. Second, it would enable critical geopolitics to investigate how different types of social movements challenge state-centred notions of hegemony, consent and power and the contestation of the "political" by the state. (Routledge 1996:509)

The role of the Antarctic and Southern Ocean Coalition (ASOC) is an important example of a multi-national ecological coalition that operates independently of member states of the Antarctic Treaty System (ATS). ASOC was created in 1977–8 and has since emerged as the most important non-state actor in the Antarctic Treaty System (Darby 1994:49). Its political strength lies in its capacity to relay information about environmental issues swiftly and effectively to interested parties. There are three major ASOC centres, based in Sydney, Washington and Wellington, which form the basis for a quasi-secretariat. ASOC has been very effective in publicizing environmental affairs in the 1990s and demonstrated that polar affairs are not, therefore, exclusively defined by nation-states and the ATS. In recognition of that fact, the ATS granted ASOC the formal status of observer within the meetings of the ATS (Darby 1994:52).

More generally, economic, informational and political globalization are contributing to the composition of a world where states are just one site of flows and connections rather than the containers of political power and action (Luke 1996). Whilst concerned with the historical and political involvement of southern hemispheric nations in the Antarctic region, it should be apparent that the international politics of the region is not encompassed by a select number of nation-states within the ATS. The debates and controversies associated with the environmental protection of the Antarctic and common heritage in the 1980s and 1990s have demonstrated the complexities of polar political conditions and the need to take other sets of political actors seriously.

Geo-graphing Antarctica

The significance of critical geopolitics and the concept of geo-graphing in relation to the Antarctic lies in several directions. Human understanding of the Antarctic as a particular kind of place has changed over the course of the present century (Manzoni and Pagini 1994, 1996; Dodds 1997). Different technologies of exploration and mapping, for example, played a key role in transforming the maps of Antarctica from the vague outline of a continent to a place filled with details of the interior and the possible distribution of resources. The collection of practical geographical information was an essential element not only in the representations of the continent and the surrounding oceans but also provided knowledge vital to the administration and mastery of space. The debates over common heritage and world park proposals in the 1980s, for example, were based on different representations of Antarctica as a geographical place. These varied representations of place can also produce differing expectations about behaviour and political management (cf. Cresswell 1996). If the Antarctic had been defined as a place of "common heritage", then the existing definitions of the Antarctic, as a place of peaceful scientific activity as managed by the Antarctic Treaty System would have been overturned. The Antarctic as "common heritage" would have raised issues such as new forms of political management, tourism, resource sharing and North–South relations. Differing policy priorities would have been advanced on the basis of changing representations of place and landscape. The debates in the 1980s on the Antarctic illustrate how particular geographical representations of the Antarctic favoured by the ATS (ie as a continent for peace and science) were challenged and transgressed by non-state actors and other sections of the international community. Place is always infected with a kind of socio-political expectation.

Five major geographical representations have been identified which broadly coincide with different periods of the present century:

(a) Antarctica as a partially filled space (1900–40)
(b) Antarctica and the Cold War (1945–60)
(c) Antarctica as a place for science (1960s)
(d) Antarctica as a place of resource potential (1970–80s)
(e) Antarctica and the global environmental challenge (1990s).

The dominant spatial representations of the Antarctic change and are significant because they have often shaped the international politics of the region and the subsequent actions of *inter alia* southern hemispheric states in the polar region. The following sections are meant to complement existing studies of the Antarctic, emphasizing the importance of political factors and structural forces affecting the world-economy throughout the present century (Hall 1986, Chaturvedi 1996). Moreover, the subsequent analyses of the SORS are then located within the broad and changing representations of the Antarctic and Southern Ocean.

Antarctica as a partially filled space (1900–40)

The first landings on the Antarctic continent coincided with an extraordinary expansion of the European and American desire to conquer nature in the late part of the 19th century. Recent innovations such as air powered flight, electricity and communication technologies had fired the imaginations of many individuals and nation-states (Kearns 1984, Wohl 1994). One who glimpsed the possibilities offered by aviation was the young Benito Mussolini. In an article for an Italian newspaper, he noted how flying would encourage "Movement towards the icy solitude of the poles and towards the virgin peaks of the mountains, movement towards the stars and towards the depth of the seas . . . Movement everywhere and acceleration in the rhythm of our lives" (Mussolini 1909, cited in Wohl 1994:287). Admittedly, the first southern polar flights were not to take place until the late 1920s, but the sentiments of that dreamer were merely reflecting wider notions that modern technologies would be capable of taking possession of the globe and enabling the deciphering of the earth's mysteries (Cosgrove 1994, Dodds 1997).

The significance of late 19th century Antarctic exploration lies in the widely recognized point that this was one of the last pockets of "unclaimed and unstated space" (Ó Tuathail 1996:15). In a somewhat Conradian vein, the delegates to the 1896 International Geographical Congress had complained of the lack of reliable geographical knowledge about the Antarctic (Mill 1896, 1901). Polar authorities such as John Murray had even suggested that: "I always feel a little shame that civilised man, living on this small planet – a very small globe – . . . has not yet fully explored the whole of this little area; it seems a reproach upon the enterprise, civilisation, and condition of knowledge of the human race" (Murray 1899 cited in Baughman 1994:12). Murray's plea for polar exploration came at a time of unprecedented European spatial expansionism and imperial aggrandizement. At the same time, however, the Swedish writer Rudolph Kjellen first coined the term "geopolitics" and thus initiated an approach to international affairs which drew upon the intellectual legacies of neo-Lamarckian biology and social Darwinism to construct readings of imperial rivalries, geographical conditions and strategic advantage. The writings of the British geographer and educationalist Halford Mackinder were indicative of a more general desire to produce readings of global space in the face of a geopolitical uncertain *fin de siècle*. The appeal of those earliest geopolitical writings was also recognized by imperial administrators such as Leopold Amery who later noted in 1904 that: "it is always enormously interesting if we can occasionally get away from the details of the everyday politics and try to see things as a whole, and this is what Mr Mackinder's most stimulating lecture has done for us tonight" (cited in Ó Tuathail 1996:30). This capacity to visualize the "whole" was later to be employed by Amery in the 1920s to articulate a vision of the Antarctic continent under complete British imperial rule.

Spatial representations of the Antarctic in the early part of the present century were shaped by the precious knowledge gleaned over centuries of exploration and commercial activities and by the imagination of explorers, imperial administrators and sailors. In the early 20th century, usually privately funded explorers such as Scott, Charcot, Mawson and Shackleton extended the discoveries of 19th-century adventurers (Beck 1983). With the added knowledge that whale oil was likely to be a valuable commodity in the inter-war period, the British state set about the political and commercial annexation of the Antarctic. The figure of Leopold Amery, as an Under-Secretary of State at the Colonial Office, looms large in the decision to pursue a policy which would enable "the whole of the Antarctic [to] be included within the British Empire, and that, while the time has not arrived that a claim to all continental territories should be put forward publicly, a definite and consistent policy should be followed of extending and asserting British control with the object of ultimately making it complete" (British Archives, Colonial Office 1920). The imperial vision of the alpine-loving Amery was simply breathtaking but not unusual: the Antarctic continent, for imperialists and explorers alike, was an unexplored space that had to be known (through the promotion of expeditions) but also possessed politically (through annexation), economically (through the control of whaling) and epistemologically (understood within western scientific thought).

This geographical depiction of the Antarctic was quite different from Western representations of the Southern Ocean. Through a combination of whaling and sealing activities, acts of geographical discovery, the experiences of the First World War around the Falkland Islands and an appreciation of the importance of sea lanes, the Southern Oceans had been charted and mapped far more thoroughly than the Antarctic continent. Indeed, the British Admiralty implicitly reflected this geographical state of affairs when it announced in 1914 that henceforth the term "Southern Oceans" would not only include the ocean directly south of Australia but would also refer to the oceans below South America, South Africa and New Zealand (British Archives, Colonial Office 1919). The cartographic empiricism of the

British Admiralty was indicative of this period: the Southern Ocean was depicted as a flat and uniform surface divided into whaling routes and sea lanes coupled with maps of the Antarctic depicting previous episodes of exploration and the residual blank spaces in need of completion (Dodds 1997: 50–2).

Prior to the Second World War, one dominant intellectual movement in the West was neo-Lamarckianism, more commonly referred to as social Darwinism (Livingstone 1992). The natural character of states was said to be influenced by biological "needs" which had to be satisfied in a competitive Darwinian world. The Antarctic continent was interpreted by some scholars such as Griffith Taylor as being a kind of "test" of human fitness and environmental flexibility (Taylor 1920, Hayes 1928). More substantially, the varied claims made on the Antarctic continent between 1907 and 1943 were underwritten by appeals to natural boundaries and the needs of the organic imperial state. Natural features were used to designate the "natural" area of a state. The development of sector theory to justify those claims, initially in the Canadian Arctic in 1909, reflected a widespread belief that lines on a map conveyed the impression that a state had a right to territory either on the basis of prior episodes of exploration or that geographical proximity justified the state's claim to further territories. Argentine writers such as Admiral Storni in *Intereses Argentinos en el Mar* argued that the Argentine nation should look to the sea as a source of unity in spirit and purpose. He wrote that: "In this great common attractive force resides one of the real and permanent causes of national union, it makes geographical unity and in the end political unity" (Storni 1967:30). As we shall see later modern Argentine geopolitical thought argues that the Argentine nation has wider territorial interests than Storni's work suggested. The point remains, however, that notions such as the state as an organism were powerful metaphors which enabled territorial claims to the South Pole and elsewhere (Hepple 1992). The Swedish writer, Kjellen argued against Norway's independence from Sweden on the basis that the Scandinavian mountains did not constitute a natural boundary (Holdar 1992:315).

Within Western political discourse, the dominant spatial representations of the Antarctic were shaped by a naturalized geopolitics (Agnew and Corbridge 1995:56–65).The declarations of national sectorial claims to the Antarctic were reflective of these forms of naturalized geopolitical reasoning. As one authority on international law has noted:

> Sector claims advanced in the south polar region have rested on formulations of discovery, as well as on various combinations of appropriations, geographical propinquity, and acts of administration that have been asserted . . . no natural baselines are available for constructing sectors [in the southern oceans]. Those lines that have been drawn are merely artificial demarcations, perhaps done for political purposes, but lacking adequate legal rationale under international law

> . . . Sectorization has been principally viewed as a convenient political device that is employed by some states to rationalise their national policy positions in the polar regions. (Joyner 1992:56–7 and 58)

Despite the reluctance of international lawyers to give credit to the legal veracity of geographical proximity, it has been interpreted as a convincing measure of sovereignty in many parts of Europe, North America and South America.

The intellectual and epistemological assumptions of geopolitical discourse had a substantial impact on modern representations of places including the Antarctic. The claims to nomothetic knowledge (rather than idiographic) depended upon the separation of the scientific from a subject position. It is fair to argue, as Peter Taylor has done in the past, that in retrospect the positionalities of various authors is not difficult to detect: Mackinder was obviously writing from a particular British and imperial point of view (Taylor 1993). However, the separation was important as Agnew and Corbridge (1995:57) acknowledge, because "even as a particular 'national interest' was self-evidently addressed, this was framed by a perspective in the realm of nature rather than politics". Further, the visualization of the world into particular political and economic regions was usually attributed to natural and innocent forces rather than as a consequence of ideology or politics. The prefiguring textuality of the geopolitical order and its associated mapping, naming and citing of spaces was not really acknowledged by writers such as Mackinder or Ratzel (see Ó Tuathail 1996:23). Finally, as with other contemporary writers such as Joseph Conrad, the world beyond the immediately familiar was often depicted as blank and/or empty but was then subsequently filled and labelled according to the scales and measures of Linnaean-inspired science of natural history (Pratt 1992). The world was then known and possessed not only politically but also in an epistemological fashion. The imperial expansion of the European state system gave rise to more aggressive forms of geographical colonization and knowledge collection.

The "natural" character of states was fashioned by their so-called "natural" boundaries. That implied, of course, that existing boundaries were not necessarily historically and politically representative of the state as an organism. It also implied that the natural area of a state could be defined using natural features rather than social or political criteria. Mark Bassin's work on the construction of geographical space in the late 19th century illustrates well how Russian intellectuals and geographical societies argued that Russia was a continental nation that straddled Europe and Asia (Bassin 1991, 1994). In the present century, South American nations such as Argentina have maintained that their natural boundaries stretch over thousands of miles of sea, ice and space. Unsurprisingly, the German concept of *Lebensraum*, with its roots in the notion of naturalized cultural boundaries, has been important in Argentina and Chile (see Child 1988).

The Antarctic continent was often depicted as a place that possessed no natural boundaries and hence imperial powers and post-colonial states alike proceeded to devise lines on the blank map. These designations continue to haunt the polar continent to this day.

Antarctica and the Cold War (1945–60)

The origins of the Cold War lie with American attempts to rebuild Western Europe and Soviet attempts to secure a zone of influence in Eastern Europe (Mason 1991). While there remain many competing explanations of the Cold War, there can be little doubt that the initial phase of the conflict was "reinforced by a crisis here, a Soviet move there, and an analysis of the protagonists which insisted that Moscow was impelled to expand and that only the United States could prevent it from achieving world domination" (Cox 1990:30). Both the United States and the Soviet Union possessed global ideologies of domination and/or influence. Geopolitical space became compartmentalized in terms of a twofold distinction, itself based on old distinctions of traditional and modern and a new one between ideological and free (Escobar 1995). Actual places only became meaningful once they were slotted into these geographical categories (Agnew and Corbridge 1995). Thus the polar regions were placed within the discursive economy of the Cold War; and peoples, regions and cultures were transformed into a militarized geography in which acronyms such as DEW (Distant Early Warning) filled the maps of the North Pole. Accordingly, the varied features of its physical environment – fjords, oceans, bays, inlets, mountain ranges and so on – were judged "likely to be involved in a major way in an armed conflict between the great powers" (Jalonen 1988:157).

One of the most remarkable aspects of the Cold War was that the Antarctic continent was not of more pressing concern to the superpowers. The Argentine diplomat Roberto Guyer argued that, by the 1950s, geopolitical understandings of the Antarctic had altered because "the presence of the United States and the Soviet Union, with the vast potential of their resources, now gives Antarctica the clear character of a geopolitical unity. It is no longer dealt with piecemeal in sectors, but as a whole" (Guyer 1983:268). Given the militarization of the Arctic, it was, however, still possible for the United States and the Soviet Union to agree to discuss the future politics of the region throughout the mid- to late 1950s (Beck 1986, Chaturvedi 1996). Although many scholars have quite reasonably noted that the superpowers and their respective allies managed to keep the continent out of the Cold War contest, it would be unreasonable either to argue that the spatial representations of the Antarctic were unaffected by Cold War discourses or that political factors were unimportant in determining the policies of the superpowers.

To illustrate this point, it is necessary to provide an element of political context which was quite specific to the Antarctic and South Atlantic region. The 1940s witnessed growing tension between Britain, Argentina and Chile over territorial claims to the Antarctic peninsula (Beck 1986, Dodds 1996b, 1996c). In the aftermath of the Second World War Britain had established a Falkland Islands Dependencies Survey (FIDS) to protect and enhance their claims to the region, whilst the Perón government had been equally active in promoting Argentine expeditions to the region in support of a formal claim to an Antartida Argentina. President Gabriel Videla of Chile had visited his country's bases in the Chilean Antarctic Territory in 1948 and had promised to support further endeavours (Mericq 1988). It is well appreciated by most polar historians that the British and the Argentines clashed in various parts of the Antarctic peninsula (the most notorious being in Hope Bay in 1952). From an American point of view, the "scramble for territory" between the three parties created fears that the Soviet Union might intervene in Antarctic affairs and make a territorial claim on the basis of past exploration. The USA was further undecided whether to make a formal claim to territory even though there is much evidence to suggest that the US State Department had made elaborate plans for a possible claim (Illingworth 1953, Plott 1969, Klotz 1990).

This uncertainty over the possible strategy of the superpowers was reflected in both political and academic circles. American expeditions like *Operation Highjump* (1946–7) reflected an expanding interest in the Antarctic when nearly 5000 men and 12 ships were dispatched "south" for polar training (Sullivan 1957:173–4). The motivation for this expedition was almost certainly a belief that the Antarctic would be a suitable arena for training purposes and one that avoided the geographically sensitive region of the "high Arctic". The Americans had earlier conducted *Operation Nanook* in Alaska with the expressed purpose of testing machinery, training men in polar conditions and assessing the feasibility of ice runway construction. There was, however, also much speculation that the Antarctic might contain strategic minerals such as uranium. After 1945–6, the US State Department reviewed its Antarctic policies on a yearly basis with the intent of ensuring that the territorial problems of the Antarctic should not *inter alia* impede scientific co-operation and exploration.

In contrast to the common academic and political assertion that "The Cold War never reached the Antarctic", geopolitical discourse at that time undoubtedly shaped spatial representations of the continent (Beck 1990c:193). Illustrations for such a proposition lie in a wide range of publications from specialist academic journals to popular magazines such as *The New Commonwealth* and even the pronouncements of political leaders. American writers were particularly prominent in their representations of the Antarctic within the matrix of the Cold War (Illingworth 1953, Scholes 1954). The *New York Times* of 21 September 1959 proposed that it was in the national interest to "ensure that this vast region should not be turned by the Russians into a kind of Antarctic Albania . . . Launching pads could be used by the USSR to blackmail the entire Southern Hemisphere from here." In this vein, the Antarctic was interpreted on the basis of its strategic importance and whether Russia might exploit its "emptiness" for its own sectional interests. This point of view was reiterated most dramatically by a study entitled "Red threat from Antarctica", found in the archives of the US State Department, published in 1959 by the magazine *Missile and Rockets*:

> At the frozen bottom of the earth Russia is moving into a position from which its missile squadrons could outflank the free world. Half of Antarctica is rapidly turning from white to red . . . From the snowy Antarctic coasts and plateau Russia is in a position to have the entire Southern Hemisphere within easy missile range . . . The earliest Antarctic threat involves the use of missile-launching conventional submarines operating out of the Soviet Union's increasing number of Antarctic bases . . . The Soviets can put their ICBM's on Antarctica itself . . . New York, Washington, Chicago, Denver, Seattle, Los Angeles . . . would be reachable . . . South America and Africa would become much more sensitive to Soviet pressure as the shadow of Soviet missiles might spread across both continents . . . Meanwhile, Soviet tractors bearing red flags are moving through the Antarctic snows in increasing numbers. (cited in Beck 1990c:198)

Discursively speaking, the above quotation is not unusual for the period as even respected journalists such as Walter Sullivan argued that: "Its vastness [Antarctica] provides a sanctuary from which aircraft could dominate the waters that, apart from the vulnerable Panama and Suez Canals, provide the only ready link between the Atlantic and Pacific and the Atlantic and Indian Oceans" (Sullivan 1957:357). The exaggerated sense of vulnerability and the invocation of "self-defence" was an important means of mobilizing public opinion on behalf of foreign ventures. The Antarctic in that sense was not immune from these pressures. The arrival of nuclear weapons and long-range flight led to a heightened sense of connection with other parts of the world (Kato 1993). One important consequence of this sense of vulnerability was a systematic exaggeration of the threat from the Soviet Union. This could range from the over-estimation of Soviet defence spending to the belief that the Soviets were hell-bent on conquering parts of the Antarctic. The Australian media too were particularly alarmed by the intentions of the Soviet polar programme in the mid-1950s (Dodds 1996b).

The most remarkable feature of this period was that whilst English-language reports on the Antarctic were filled with dire strategic predictions and fears of Soviet expansionism, 12 states were participating in a series of meetings organized by the US State Department in order to remove tension and strife from the international politics of the Antarctic. The International Geophysical Year (1957–8) is usually interpreted as an important stage of the process of negotiation and consensus for a management system for the continent (Gould 1958, Fogg 1992). The growing importance of science and scientific discourses was evident by the late 1950s as spectacular events such as the Trans-Antarctic Expedition (1957–8) led by Hillary and Fuchs seemed to indicate a high point in modernist polar achievement as explorers and their machines successfully crossed the polar continent (Dodds 1997). Notwithstanding these technological achievements, the co-existence of geopolitical and scientific discourses could often be awkward as Henry Dater of the US State Department noted in July 1959: "Because of its leadership in the Free World, it is evident that the United States could not now withdraw from the Antarctic . . . Antarctica simply cannot be separated from the global matrix. Science is the shield behind which these activities are carried out" (cited in Beck 1986:64–5).

The decision to hold the International Geophysical Year (IGY) in 1957–8 is indicative of the tensions between scientific endeavours and political concerns. There appeared to be good reasons for holding an IGY with a special polar programme. New raw data were needed given that the last concerted data collection exercise was carried out in the Second International Polar Year in 1932–3. Unusual solar activity was predicted to occur in 1957–8 and new information was wanted on the earth's atmosphere in order to help the development of modern communication systems. The Comité Spécial de l'Année Geophysique Internationale (CSAGI) gathered in Paris in May 1955 to discuss the format of the proposed IGY. Yet the Argentine and Chilean delegations, headed by their ambassadors in Paris, demanded reassurances given the impact that freedom of scientific investigation might have on territorial and resource rights to the region, whilst the Australian government was concerned, for instance, about the Soviet's intentions to locate their IGY bases in the AAT. As a compromise, the CSAGI later agreed to issue a statement noting that the activities of the IGY should have no adverse impact on the existing territorial status quo (Hall 1986).

Throughout the IGY, the American government was deeply concerned over the intentions of the Soviet Union after the ending of this research programme in December 1958 (Plott 1969). More generally, it should be remembered that the IGY was dominated by the race between the Americans and Soviets to launch a space satellite and by plans to conduct military-inspired geophysical research into weather control. Initial proposals to extend the Antarctic part of the IGY by another year were rejected by countries such as Australia and South Africa on the basis of cost and the delay in processing scientific data. However, more importantly, claimant states such as Australia and Argentina were anxious to remove the Soviet Union from the Antarctic and wanted the latter to dismantle its IGY polar bases. The decision to convene a series of informal conversations, led by the American representative Paul Daniels, over the future of the Antarctic was motivated not by a desire to further international scientific co-operation but by the realization that the Soviet Union would remain in the Antarctic post-IGY. Moreover, the launching of Sputnik I in October 1957 coincided with established American interest in the Soviets' technical ability in advanced intercontinental missile technology, space research, geophysical research into the upper atmosphere and long-range bomber capacity. For most of America's Western allies, concern over the intention of the USSR was of fundamental importance; so much so that Britain held a series of secret meetings with Argentina about their mutual security worries.

The increasingly widespread fear that Antarctica might be a possible site of Cold War tension led 11 countries (Argentina, Australia, Belgium, Chile, France, Japan, New Zealand, Norway, South Africa, the Soviet Union, the United Kingdom) to join the American government to attend a conference on the future management of the Antarctic. The secret details of the conference held in Washington DC between October and December 1959 need not concern us here (Beck 1986, 1990c). Crucially, however, in terms of the changing spatial representations of the Antarctic, the conference signalled an important moment as diplomats attempted to change the political status of the continent and agree upon a suitable geographical definition of the Antarctic. The defintion of the Antarctic region was eventually settled in 1958 using the 60 degrees South latitude. There were profound fears that the continent might either be a site of further conflict over established territorial claims or, worse still, might be subjected to the dangers posed by atomic weapons. It was reported, for example, that the Americans had detonated three hydrogen bombs over a part of the South Atlantic in August 1958 in order to investigate upper atmospheric change. This development certainly contributed to widespread southern hemispheric fears over nuclear activity in the southern seas and Antarctica. As the Argentine diplomat Roberto Guyer reminded readers: "We were thus [in 1958–9] on the threshold of a conflict which could have had serious political consequences . . . It must be remembered that these events occurred during the Cold War. We were on the verge of a situation which could have easily resulted in a confrontation between states" (Guyer 1983: 270). The construction and eventual ratification of an Antarctic Treaty in 1961 was the result of months of diplomatic endeavour and hard bargaining over the demilitarization and denuclearization of the continent. Territorial claims were "frozen" and states were encouraged to pursue international scientific co-operation. Twelve nations had created an Antarctic Treaty System which would support a new management structure, based on consultative party meetings and the exchange of scientific information, for the polar continent. However, these achievements need to be assessed against the fact that neither sovereignty issues nor resource rights were resolved by the negotiating parties.

Antarctica – zone of peace and science (1960s)

The international significance of the Antarctic Treaty was swiftly recognized by contemporary reporters and political observers (Candioti 1961, Sullivan 1961). For the first time, and at the height of the Cold War, a number of states had agreed to demilitarize and denuclearize an entire continent. Indeed, the preamble of the Antarctic Treaty urged that national interests should be subsumed to a broader logic: "It is in the interest of all mankind that Antarctica shall continue forever to be used exclusively for peaceful purposes and shall not become the scene or object of international discord." For many scholars, the old geopolitics of formal imperialism and territorial aggrandizement had, in December 1959, been banished from the foreign-policy scripts of interested member states (Chaturvedi 1994). After formal ratification in 1961, the language of science and scientific endeavour, under Article IX of the Treaty, became the new language of the Antarctic specialist and interested policy maker. The discursive transition implied in such analyses was never that obvious, however. South American geopolitical writings continued to analyse the Antarctic in much the same terms that American political discourse had interpreted the strategic significance of the Antarctic in the mid-1950s (Dodds 1993). Other southern hemispheric nations such as Australia and New Zealand never dropped their stated geopolitical and strategic concerns for the region.

The Eisenhower administration was also acutely aware that the attempts to depoliticize the Antarctic were extremely difficult in the context of a dominant Cold War geopolitical discourse which assigned places on the basis of their strategic location and peoples on their potential ideological orientation. The Antarctic was another place where the United States had to ensure that they took: "advantage of present and possible future uses of Antarctica that benefit the security and welfare of the United States and to prevent any use of Antarctica which would be detrimental thereto" (J. Mooney, US official 1961, cited in Beck 1990c:210). It would be naive to presume that the creation of the Antarctic Treaty System led to the complete political transformation of the continent in favour of scientific investigation and peaceful co-operation, but it was appreciated by many political leaders and some commentators that science, politics and technology were closely interrelated (Taubenfield 1961, Hall 1986). Research programmes on oceanography or the upper atmosphere had implications for the navigation and communication systems used by various armed forces. The United States Department of Defence was a key source of such studies including Project Argus which was designed to investigate the possibility of inducing climatic change and the distribution of the aurorae. The latter was perceived to be a major influence on communication systems. South American states, for example, continued to use their armed forces to support scientific programmes and geological research which could be used to assess the resource potential of the region.

The image of the Antarctic continent as a place for scientific co-operation and international co-operation was particularly striking within the Western scientific community (Lewis 1965, Lewis and Smith 1973). Article IV of the Antarctic Treaty was considered axiomatic to this development (Heap 1983, 1988). Henceforth, many scientists argued that the Antarctic was to be considered a unique "laboratory" for science rather than a continent characterized by the selfish pursuit of territorial claims. According to Walter Sullivan, "The Antarctic Treaty has set an entire continent aside as a laboratory available to research by all and open to inspection by signatories who seek assurance that its provisions against military activity and nuclear testing are being adhered to" (Sullivan 1961:xi). Herr and Hall have rightly noted that whilst "not all science pursued in the region has been 'science for science's sake' ", the metaphor of the "laboratory" has frequently been applied by scientists and political observers to describe the Antarctic in the 1950s and 1960s (Herr and Hall 1990:17). Its significance lies in the way that the term "laboratory" was employed to create a specific place as a spectacle or an experiment, which was in turn designed to generate consent amongst the participants of the ATS. The conventions and procedures of the Antarctic Treaty parties were important in establishing controls on actions within that space and in establishing criteria for what counted as proper scientific knowledge (Latour 1987).

The role of the Scientific Committee on Antarctic Research (SCAR) in establishing the Antarctic as a place of study and research is a neglected theme in polar studies because most of the accounts of Antarctic science have been characterized by a celebration of polar science and the explanatory power of science (Quilty 1990). Initially created in the middle of the International Geophysical Year by the International Council of Scientific Unions, the Special Committee on Antarctic Research (renamed the Scientific Committee after 1961) played a part in establishing the means by which the Antarctic was measured and understood. The preamble of SCAR's constitution reiterated this dimension: "[it is] charged with furthering the co-ordination of scientific activity in the Antarctic, with a view to framing a scientific programme of circumpolar scope and significance" (cited in Darby 1994:36–7). SCAR was intended, therefore, to act as a catalyst for international scientific co-operation; at the same time it was also hoped that individual members of SCAR would be sufficiently independent of their respective governments (Beck 1986). However, it has since been admitted by SCAR members that freedom to pursue scientific investigation has been at times compromised by the demands of national governments and rather than being independent from the ATS, SCAR has often acted like a close partner to the ATS membership (Hall 1986, Chaturvedi 1994).

Ultimately, the depiction of the Antarctic as either a "laboratory of science" or a "continent for peace and science" rested on a naive understanding of science and a lack of understanding as to the political and strategic dimensions of scientific research. Richard Laws' comments are indicative of this limited understanding of the political significance of Antarctic science: "Scientific activities are relatively non-controversial and had been demonstrated during the International Geophysical Year to defuse sovereignty issues" (Laws 1987:250). In contrast, Taubenfield concluded in 1961 that: "scientific activities have themselves political overtones . . . whatever advances science furthers strategic objectives: a station useful for gaining knowledge of our environment is ultimately strategically important by its very nature" (Taubenfield 1961 cited in Herr and Hall 1990:17). By the 1970s, new international agendas were creating opportunities for a fundamental reassessment of the Antarctic as a particular kind of place. The Antarctic was rapidly transformed by the international community from a "continent of science" into a place overtly invested with resources and mineral potential.

Antarctica as a place of resource potential (1970s–80s)

In the 1960s, commercial fishing in the Southern Ocean became increasingly important due to the legal developments relating to the declaration of Exclusive Economic Zones (EEZs). The demarcation of EEZs effectively: "barred the Soviet and other deep-water fleets from many of their traditional fishing grounds causing them to look southwards" (Vogler 1995:77). In earlier phases of Antarctic resource exploitation, whaling was of primary importance. In the post-war period, the search for krill and fin fish such as Antarctic cod stimulated a considerable investment in fishing fleets from states such as the Soviet Union, South Korea and Japan. At the same time, increasing numbers of academics and policy makers were predicting that the world's food supplies might not match rising population levels. The proposed exploitation of these living resources in the Southern Ocean forced the ATS in the 1970s to consider new resource conventions in the light of earlier tragedies of the over-exploitation of seals and whales (Holdgate 1987, Miller 1991a).

The resource potential of the Antarctic and Southern Ocean was formally raised by the British delegation as an issue due formal consideration at the Tokyo ATCP meeting held in 1970. After a degree of reluctance, the ATCP agreed to place the issue on the agenda for their next meeting under the heading: "Antarctic resources: effects of mineral exploitation". The agenda item was not discussed in 1972 and delegates agreed to postpone discussion until the following ATCP meeting, eventually held in Oslo in 1975. However, the Conservation of Antarctic Seals was negotiated in 1972 and this provided regulatory mechanisms for the exploitation of that living resource. The Oslo meeting, however, concluded that the remaining resource question would have to be split into two parts: living and non-living (ATCP 1975). It was decided that a moratorium be established on the latter so that the ATCP could first concentrate on developing a living resources regime. The delays in establishing a minerals regime also suited nations such as Argentina and Australia, which concluded that their nations were ill-prepared for the possible exploitation of the Antarctic in comparison to Japan or the United States. The significance of this development has also to be interpreted within the wider setting of the 1973–4 oil crisis, debates over a New International Economic Order, North–South relations and neo-Malthusian fears that population growth would outstrip the world's resources (Lovering and Prescott 1979, Peterson 1980, 1988, Schroff 1982).

From a geographical point of view, this period increasingly witnessed the transformation of the Antarctic from a place for scientific endeavour to a place of common economic heritage. The region became increasingly entangled with the new international agendas of the 1970s and the preoccupation with resource potential in an increasingly inter-dependent world (Peterson 1980). Simultaneously, the ecological movements, led by Greenpeace, were beginning to focus their attention on the polar continent in the late 1970s. The writings of Keith Suter were indicative of a mounting concern that the Antarctic should be declared a "world park" rather than become an object of resource exploitation (Suter 1980). The Sri Lankan delegate at the Law of the Sea Conference had, for example, proposed in 1975 that Antarctica should be considered a common heritage of mankind and therefore considered in the same terms as the deep seabed (Vogler 1995:81). The declaration of a Convention on the Conservation of Antarctic Marine Living Resources (CCAMLR) in 1980 was a significant development in the face of these international pressures. The Convention established some important principles of "rational use" of marine resources which would enable the "ecological relationship between harvested, dependent and related populations" (Vogler 1995:85). Notwithstanding some controversies over membership to the Convention, this agreement signified a shift in the dominant geographical representation of the Antarctic from a "scientific laboratory" to a resource-rich continent.

During these negotiations the SORS were reluctant to allow their sub-Antarctic possessions to be included in the provisions of the CCAMLR. Australia (Heard Island) and South Africa (Prince Edward Islands) argued in conjunction with France (Kerguelen and Crozet) that their fishing rights should not be compromised by the Convention. The declaration of exclusive fishing zones by Australia and South Africa in 1979 reflected their fears that these resource rights could be compromised either by a new convention or by other fishing nations. The resource potential of the Southern Ocean increasingly influenced SORS policies towards the region and explains their eagerness to support oceanographic and fishing-based research. The inclusion of Article IV in the final convention enabled a settlement of conflicting jurisdictions through the declaration of "coastal state jurisdiction". The provisions of Article IV allowed the claimant states to protect their sovereign interests whilst at the same time non-claimant states did not have to recognize the coastal state jurisdiction of those claimant states (Chaturvedi 1990:133). In doing so, the ATS sidelined the controversial issues of sovereignty both north and south of the 60 degrees boundary.

This geographical transformation was codified still further with the ongoing discussions over the Convention on the Regulation of Antarctic Mineral Resource Activities (CRAMRA). Speculation over the existence of oil, mineral and natural gas resources reached a high point in the late 1970s and early 1980s (Shapley 1983, 1985). In 1982, the ATCP began negotiations over a minerals regime which was designed to regulate the exploration and exploitation of resources (Cook 1990). The Wellington Convention was eventually rejected by France and Australia in 1988–9, after six years of delicate and painstaking negotiations (Kyriak 1996). The campaign against CRAMRA by ecological movements such as Greenpeace and the Antarctic and Southern Ocean Coalition played an important part in altering official Australian and French positions towards the Convention. The establishment of an Antarctic base, World Park Base, coupled with demands for the Antarctic to be declared a World Park or Global Wilderness had the effect of raising public opposition to CRAMRA in Australia, New Zealand and South Africa. Moreover, Greenpeace's exposure of Antarctic waste dumps close to existing bases stimulated public awareness of the importance of polar environmental conservation. The journalist Michael Cross has argued that: "Governments tend to deny that they change their policies because of Greenpeace's actions. But the organisation must deserve some credit for its part in creating, virtually from scratch, a public lobby that has caused governments to take notice of the Antarctic environment and to move towards the world park" (Cross 1991:32). Among the United Nations, a number of Third World nations, led by Malaysia, declared that the Antarctic should be understood as a common heritage rather than as the managerial and political preserve of a select number of nations. In a rather different vein to the ecological movements, the UN General Assembly challenged the intellectual hegemony of the ATS throughout the 1980s (Beck 1986, Kyriak 1992). The existing members of the ATS, Australia, in particular reasserted the intellectual and political hegemony of the treaty system. As mentioned before, this rearguard action was not just over principles of political representation (ie should the United Nations play a stronger role?) but also over geographical representation (ie how should the Antarctic be understood as a particular kind of place?). The rejection of CRAMRA and the ratification of the

Protocol in 1991 were fundamental in restoring the moral and political authority of the ATCP.

Antarctica and the global environmental challenge (1990s)

During the 1990s, Western social science has sought to advocate the global environment as an urgent priority for the international political system (Hurrell 1994, Chatterjee and Finger 1995). Unlike those earlier debates in the 1970s, the ending of the Cold War renewed hope that environmental and developmental issues would now take priority over strategic and geopolitical considerations (Kliot and Waterman 1992). It has been subsequently argued that the concept of security will have to be redefined or reconceptualized in order to take account of the trans-boundary nature of the environmental challenge rather than the military dimension (Unger 1991, Dalby 1993). Sociologists such as Ulrich Beck have argued that these kinds of concerns about the environment are emblematic of late modernity and the onset of "risk societies" (Beck 1992). He suggested that Western societies are confronted by a double bind: a growth of terrifying ecological risks has been combined with an associated lack of faith in science's ability to secure solutions to these environmental problems. Beck reasoned that these societies are more concerned with ecological security – although this is considered problematic because of the growing uncertainty about the future – rather than the creation of material wealth. In the aftermath of the Chernobyl disaster, the failure of the Soviet Union or the European Union states to deal adequately either with immediate environmental degradation or possible long-term environmental change has given added weight to these concerns (Hurrell 1994, 1995).

During the 1990s, the Antarctic continent has increasingly been represented in global ecological terms. The mounting body of evidence purporting to link the polar continent to possible global climatic change has been used by scientists and scientific bodies to explain and legitimize their endeavours. In 1991, the then director of the British Antarctic Survey reported: "I welcome the new challenges which will emerge as the locus of Antarctic affairs returns to the fundamental scientific contributions of the regions to urgent issues of global concern" (*British Antarctic Survey Annual Report 1990–1*). It is precisely because the Antarctic is perceived to be of global environmental significance that its geographical understanding and cultural significance is of great importance. On the eve of the 1992 Earth Summit, the ATS, led by Germany (which had hosted the previous ATCM in 1991), contacted the United Nations to inform the organizing panel that the Antarctic should be included in the remit of those discussions because the Antarctic was already managed by a separate political and legal system. Ultimately, the Earth Summit did not consider the future of the polar continent because of the diplomatic pressure imposed by the

ATS membership. However, it did indicate that debates over resource rights and political responsibility have not evaporated and that the Protocol has not resolved wider issues of resource sharing and global governance (Vogler 1995, Dalby 1996).

The transformation of human understanding of the Antarctic has been substantial since the turn of the present century (Dodds 1997). Whilst there are parts of the physical geography of the Antarctic that can still surprise (such as the recent discovery of a large underwater lake buried within the icecap of eastern Antarctica), remote sensing and aerial mapping have been instrumental in producing accurate maps of the region. Moreover, a combination of oceanographic, atmospheric and geological research is providing new stores of geographical knowledge about the Antarctic and its ecosystems. In comparison to those earliest reports on the polar continent, the Antarctic is now a place that is closely monitored and evaluated by a range of political, media and ecological groups. It is now common practice for ecological groups such as ASOC and Greenpeace to attend consultative meetings and submit written and oral testimonies about various environmental or administrative issues to the ATS. The journal *Eco*, for instance, was responsible for recounting stories about ATS meetings and commenting upon issues such as the slow ratification of the Protocol by the ATCP since the Madrid meeting in June 1991. Ecological movements served to press for a decision on the location of a secretariat and the liability annexe issue for the Protocol. Antarctica has become indicative of an environmental and resource globalism. Whilst debates over mining, environmental protection and common heritage in the 1980s and 1990s have been bound up with a wider concern for the management of the global commons, the controversies over ozone depletion, icecap melting and environmental pollution were mobilized by scientists and ecological groups who argued that the Antarctic was a global environmental challenge. On the eve of the launch of a new Greenpeace Antarctic expedition, a spokesperson for Greenpeace New Zealand acknowledged:

> The critical attention paid by environmental NGOs to logistical and scientific activities in Antarctica is not going to disappear, and, if anything, is likely to increase as tourism boosts the numbers of independent observers travelling to the region. However, increasingly to this will be added a greater understanding of the importance of high quality, international Antarctic research to life elsewhere on this planet. (Dalziell 1997:4)

Summary

It has been shown that the dominant spatial representations of the Antarctic have changed over this century. The reasons for these changes reside within the complicated intersections of economic transformation, geopolitics, aerial and photographic technologies, political change and geographical exploration. The subsequent analyses of the Southern Oceanic Rim States (SORS) will be located within these varied historical and geopolitical frameworks. One theme that will remain of central importance is the belief that naturalized forms of geopolitical reasoning have justified and legitimized the activities of the SORS in the Antarctic and the Southern Ocean. The future management of the Antarctic and Southern Ocean in the face of competing sovereignties and proposals for collective management remains the major problem for these states. In an era when the global commons of the earth and outer atmosphere have become increasingly politicized because of our improved ability to exploit and degrade them, old realist discourses based on national security and sovereign states have been challenged by calls for new forms of political authority and governance (Walker 1993, Archibugi and Held 1995, Vogler 1995).

Finally, this chapter has highlighted the importance of the construction and circulation of geographical knowledge about the Antarctic for the subsequent production of geopolitical readings of the polar region. Geographical knowledge is a form of power. State funding for surveying, mapping and other forms of scientific investigation played a role in enabling the SORS to maintain their territorial and maritime interests in the Antarctic and Southern Ocean. This struggle over the geography of the Antarctic is not simply a struggle over the naming of places or the charting of space but is also intimately linked to the imagination of place. The Antarctic has been a place of contestation as the SORS and other states have struggled to represent not only the physical geographies of the ice and sea but have also struggled to envisage the Antarctic as part of their national space and identity.

3

Argentina and Antartida Argentina

According to the earliest writings of Admiral Roberto Storni, Argentina has always had a natural geographical interest in the surrounding seas and the Antarctic continent (Pittman 1981, Child 1985, Hepple 1988). Argentina's claims to the Antarctic peninsula have been consistently understood within a naturalized geopolitical discourse (Agnew and Corbridge 1995:56–65). In Storni's work, *Intereses Argentinos en el Mar,* the Argentine state was conceptualized as a territorial state which had particular "needs" and "demands". The surrounding sea was said to possess a special attraction for Argentina: "In this great common attractive force resides one of the real and permanent causes of national union; it makes geographical unity and in the end political unity" (1967:22, originally published in 1916). The Argentine nation, according to his analysis, had to realize the strategic importance of the Atlantic Ocean, including the Antarctic region, and ensure that Argentina controlled resource exploitation and commercial traffic flows in and around her territorial seas (Fraga 1979, 1983, Gough 1992).

Argentina's involvement in the Antarctic and the Southern Ocean signifies an important component of the historical and political development of the Republic and compared to other SORS, Argentina has been the most hawkish in terms of protecting her territorial claims to the Argentine Antarctic sector (see Comision Naçional del Antarctico 1947, 1948, 1949). Since the mid-1940s, a formidable amount of legislation, public speeches, acts of occupation and polar base construction has been mobilized for this purpose (eg *Atlas de la Republica Argentina 1965*). As the Argentine Foreign Minister Molina declared in February 1953:

> Argentine sovereignty over the territory [Argentine Antarctic] is based on deep rooted historical rights . . . which are spiritually identified with the feelings of the

> entire people of the nation; on the superior geographical position of the Republic; on the geological continuity of its land with Antarctic territories; on the climatological influence which the neighbouring polar zones exercise on its territories . . . and finally on its uninterrupted activities in the Antarctic territory itself.

Molina's comments were also indicative of an educational system that has invested a considerable amount of energy ensuring that Argentine citizens are instructed in the geopolitical realities of the Republic (Reboratti 1982). The relative significance of subsequent geopolitical ideas such as "tri-continental Argentina" and the "Argentine Sea" was a reflection of a territorial public culture which has placed great emphasis on the integrity of Argentina and the need for vigilance at the frontiers and boundaries (Caviedes 1988, Escudé 1992).

This chapter is organized into six parts: First, the historical context is established through a brief review of Argentine activities in the Antarctic region (Fraga 1979). Second, the Argentine claim to the Antarctic is considered with particular reference to the first Perónist era (1946–55). President Juan Perón was undoubtedly a major figure in raising Argentine public interest in the Antarctic and the South Atlantic region (Perón 1948). It was under his government that the National Commission on the Antarctic published details of a formal claim on 12 March 1947 (Pittman 1981:681). The third part examines Argentina's interests in the polar region since the creation of the Antarctic Treaty System (ATS) in 1961. Argentina has never renounced her territorial interests in the Antarctic and was adamant that her sovereign rights were unaffected by the signing of the Antarctic Treaty. Argentine attitudes over the resource conventions for living and non-living resources are also considered. Plans to create a South Atlantic Treaty Organization (SATO) and to reclaim the Falklands/Malvinas are reviewed here in order to illustrate how the Antarctic was considered as one part in an interlocking representation of the polar continent and the Southern Ocean (Hurrell 1983, Dodds 1995). This appeared to be the case under the various military regimes of the 1970s and early 1980s. Moreover, as with the Chilean scientific programme, the armed forces have contributed in terms of logistical facilitation which has led to accusations from other nations such as Britain and Australia that the Argentine polar programme has more soldiers attached to it than scientists (Fraga 1979, Comision Naçional del Antarctico 1988). The controversies over mineral rights and the negotiations over CRAMRA also need to be contemplated because they were closely linked to Argentina's robust defence of the ATS in the face of pressures from other developing states in the UN General Assembly. Finally, consideration is given to Argentine interests in the Antarctic in the aftermath of democratization and new conventions such as the 1991 Protocol on Environmental Protection (Beltramino 1995).

Argentine interests in the Antarctic and South Atlantic

It could be argued that Argentina's claim to the Antarctic peninsula was inherited from Spanish imperial claims established under the Treaty of Tordesillas in 1492 (Acuna de Mones Ruiz 1948, Alzereca 1949). On the basis of the principle of *uti possidetis juris*, Argentina inherited the Spanish sphere of interest. As Jorge Leal, director of the National Antarctic Directorate noted: "Since the beginning of nationhood, because of our inheritance, we Argentines are continuing the secular actions performed by Spain . . . The Republic is determined to occupy, command and manage the farthest reaches of its territory" (*Correo del Pinguino* March 1970:11). This perception forms an important base for the formal claim to the Antarctic peninsula. Other pertinent aspects of this discussion must include the fact that Argentina has continuously occupied Laurie Island in the South Orkneys since 1904. Ironically, given the subsequent Anglo-Argentine friction over the Antarctic peninsula, the inspiration for the Argentine presence in Laurie Island came from the Scottish explorer William Spears Bruce, who led the Scottish National Antarctic Expedition between 1901 and 1904 (Speake 1996). The occupation of this base was a crucial element in the subsequent claim to an Argentina Antartida. A decree issued by President Julio Roca, at the turn of this century, also confirmed that Argentina intended to establish new bases in the "southern seas of the Republic" (cited in Alzereca 1949).

Until the eve of the Second World War, Argentine interest in the Antarctic was concentrated on the annual relief of the station at Laurie Island and the fishing activities of the South Georgia-based Companía Argentina de Pesca (Headland 1985). Occasional directives or formal memoranda from personages such as the director of Argentine Posts and Telegraphs to the director of the Universal Postal Union in September 1927 reminded wider audiences that Argentine territorial jurisdiction extended to "polar territories which have not been delimited" (cited in Bush 1982). This relatively modest commitment has to be understood within the changing politico-economic circumstances of the Argentine Republic. Trading relations with Britain after the 1932 Ottawa Conference had become increasingly strained and were declining in light of the economic circumstances. Whilst Britain could turn to its trading and political empire, the Argentine economy went into decline (Gravil 1985, Rock 1987). The Argentine presence in the South Atlantic, however, mirrored other southern hemispheric nations in terms of establishing a record of permanent occupation. A post office was created on Laurie Island, plaques declaring Argentine sovereignty were scattered throughout the island and the Argentine flag was hoisted over the station. Otherwise interest in the Antarctic itself remained minimal until 1939, even though the whaling company Companía Argentina de Pesca continued to contribute to scientific knowledge of the Southern Ocean (Headland 1985).

Then in July 1939, Argentina created a permanent interdepartmental National Antarctic Commission to co-ordinate and plan Argentine expeditions to the Antarctic and outlying islands. The motivation for such a commission was prompted by the decision of the Norwegians to press a formal claim to the Antarctic in 1939 (Hanessian 1965:11). From the Argentine perspective, the territories surrounding the Antarctic peninsula, even if under formal Argentine jurisdiction, had already been claimed by Chile, Britain and, to a lesser extent, by Norway.

The significance of the National Antarctic Commission not only lies in the political context of its creation but also in the employment of naturalized geopolitical reasoning. Under Decree Number 35,821 of 15 July 1939, the Commission was created on the basis that:

> the geographical situation of the territory on the most southern part of South America assigns to it on scientific and political grounds a natural interest in the questions arising from the study of the problems of the Antarctic continent, whose geographical resemblance to our territory, as also its probable link with this continent is, moreover, a fact which it is of interest to elucidate . . . The Committee shall make a comprehensive study of the present state of Antarctic problems and shall propose a plan of action to include questions of oceanography, meteorology, fisheries and any explorations which Argentina might undertake. (cited in *Polar Record* 1946:412)

The deployment of a geographical dimension into a presidential decree, where appropriate, is hardly controversial. This kind of geographical depiction plays a part in pushing on to the political stage issues of territorial acquisition, political conquest and occupational control (Hepple 1992:140). The subsequent pronouncements on the Antarctic, especially under the Perón government, tended to be slight in detail regarding the physical environment of the Antarctic continent in favour of territorial occupation and legal control (Paiva 1992). It was increasingly argued in the 1940s that Argentina's "manifest destiny" lay in the successful colonization of the southern lands beneath Patagonia.

As a consequence of the above presidential decree, the Argentine government issued instructions for expeditions to be sent to Graham Land (or Tierra San Martin) and to the South Shetlands in 1942 and 1943. Argentine parties were instructed to initiate surveying programmes and deposit claims to sovereignty in those places. A post office was created in the South Orkneys and the Argentine flag was hoisted over Deception Island. With the British preoccupied with fighting fascism in Europe and the Far East, the last major British expedition to the region had been the 1934–7 British Graham Land expedition. The Argentines occupied British bases and were unhindered in their progress (Escudé 1983). A formal claim to an Argentine Antarctic sector was later announced in 1943 on the premise that the Argentine Antarctic sector was an integral part of the metropolitan territory and had been so since independence in 1810 (Casellas 1981, Genest

1985). The basis for such a claim was derived from a mixture of historical precedents, recent polar activity, geographical proximity and on the geological continuation of the Andean mountains through island chains on to the polar continent. Regardless of their international legal merits, these arguments have been powerful resources from which governments have argued that Argentina's boundaries were determined by natural forces rather than by political or social criteria.

The response of the British government to Argentine claims was swift. The latter's claims to the Antarctic substantially overlapped with the existing British claim and the reluctance of the Argentine regime to declare war against Nazi Germany meant that British officials were increasingly concerned about the political sympathies of the Argentine military who had taken power in 1943 (Escudé 1983, Dieguez 1989). The British War Cabinet therefore ordered various government departments to formulate an expedition to restore British claims to the Antarctic region. *Operation Tabarin* (later renamed the Falkland Islands Dependency Survey) was organized by the navy and designed to create permanent stations in the Antarctic peninsula as well as to deter the ongoing activities of German commercial sea raiders (Beck 1986, Dodds 1996c). Political activities were limited to removing evidence of Argentine claims by the hoisting of British flags and the deposition of plaques over Graham Land. The Argentine Ministry of Foreign Affairs and Worship later responded to the British government on 20 February 1943 with the following:

> The Argentine government reaffirm on this occasion their sovereign rights over all Antarctic lands and dependencies south of 60 degrees latitude South and between meridians 25 degrees and 68 degrees longitude West . . . The Argentine government regret that the British government have considered it opportune to take such measures as the removal of the national emblem and signs on Deception Island where, previous to the placing of such, there was no effective occupation or possession belonging to the British government.

However, *Tabarin* was significant because it precipitated the development of national Antarctic programmes and renewed Anglo-Argentine competition over the Antarctic and South Atlantic (Dodds 1996a). The rationale for the British actions was explained concisely by the Ministry of Information in June 1944: "A major purpose of the expedition now is to act as a counter-measure to encroachments which the Argentine government has made in the Falkland Islands Dependencies in recent years" (cited in Dodds 1996c).

Perón and Antartida Argentina

Argentine public culture, according to the historian Carlos Escudé, has been thoroughly immersed in territorial issues, largely as a result of public education and media reporting. In a series of analyses of Argentine territorial nationalism, Escudé has argued that primary and secondary school education remains wedded to a belief that Argentina has suffered territorial "losses" at the expense of its South American neighbours and outside powers such as Great Britain (Escudé 1986, 1988, 1992; Escolar et al 1994). This apparent fixation with territorial boundaries and the state has undoubtedly affected Argentine attitudes towards the Antarctic continent. In the 1920s, for example, the first school geography textbooks began to refer obliquely to an "Argentine Antarctic territory". The creation of an Antartida Argentina in the 1940s was shot through with arguments that Argentina's territorial destiny depended upon these claims. As with the Chilean case, maps of Argentina's Antarctic interests were swiftly deployed to strengthen these claims and geopolitical ideas about the territorial state. After the Second World War, Argentina was frequently delimited on the basis of using natural features to designate the territorial area of the state.

The 1940s were a crucial decade for the Argentine Republic. Since the coup of Jose Uriburu in 1930 there had been a growing consensus amongst intellectuals and political leaders that Argentina had become increasingly peripheral in the world-economy and international political system. The sixth largest economy in 1900 had undergone a rapid decline by 1930. It was proposed, therefore, that Argentina needed a new sense of mission which would chart the passage for a return to industrial strength and political influence. In the aftermath of the inter-war depression, Argentine leaders decided that the Republic had to develop new trading and economic relations independent of Britain and the United States. Argentina's ambivalence over the Second World War, therefore, was symptomatic of this cultural and political change (Melinsky 1978, Crawley 1984). The growing diplomatic and military interest in the Antarctic and Southern Ocean was part of this reform package precipitated by the Concordancia and accelerated by the Grupo Unido de Officiales led by Colonel Juan Perón.

With the formal election of Juan Perón as president in 1946, Argentine interest in the Antarctic reached unprecedented importance in national politics. In the same year, Perón ordered the despatch of a major expedition to Tierra San Martin (Graham Land) and resurrected the Antarctic Commission. With the help of a public education programme, Perón was determined to make Argentina more "Antarctic conscious". From 1946 onwards, Argentine geography school textbooks, for example, fomally reflected Perón's belief that Argentina was composed of an Antarctic and South Atlantic dimension. School children were instructed to memorize these figures and teachers were urged to employ maps which depicted Argentine territories in the Antarctic and South Atlantic (see Escudé 1986). Carlos Escudé (1992:10) has noted how geography and history textbooks were altered in a short period of time (1938–48) to reflect the new mood in the Republic:

> In 1939, he [L. Dagnino Pastore, a popular textbook writer] wrote that Britain "possesses" more than eight million square kilometres in Antarctica (to which he applied the British term, the Falkland Islands Dependencies); in 1940 he changed the word "possess" for the expression "attributes to itself", adding that Argentina might get a part of this if the criterion for the distribution of territories used in the Arctic were [*sic*] applied; in 1944 he stated that Argentina has "unquestionable rights" and "legitimate bases for sovereignty over a vast Antarctic sector"; in 1946 he reports that Argentina has made it known to the world that it claims the Antarctic sector over which "it has rights"; and finally in 1947 he writes matter-of-factually [*sic*] of an Antarctic sector over which Argentina "exercises authority".

After 1947, the Republic had officially "expanded" from 2.8 million to 4 million square kilometres, as Alexander Land was incorporated into the Argentine Antarctic sector. The Military Geography Institute too was instructed to produce new maps which depicted the Argentine Republic as encompassing:

> River islands of the Plata . . . The Malvinas Islands . . . The islands of Lennox, Picton and adjacent islands in the Beagle Channel . . . the South Orkneys, the South Georgias, the South Sandwich Islands and the Argentine Antarctic sector which we maintain sovereignty over. The limits of the latter are: meridians 25 and 74 degrees West of Greenwich and from the parallel of 60 degrees south to the South Pole, it is to be called: Argentine Sovereign Territory in the Antarctic Sector. (cited in Escudé 1990:174)

This change of spirit was also reflected in official publications such as *The Argentine Antarctic* produced by the Argentine government which declared that:

> The Argentine Antarctic is a perfect reality. It is not a whimsical affirmation that only appears in documentary form in maps drawn up in cabinets; the documentation on which this reality is based is inscribed by acts of permanent and active occupation, and which served [*sic*] as proof of possession in all countries and according to all international laws. (International Service of Argentine Publications 1954:6)

The claims to permanence and substance need to be counter-balanced with an appreciation that the demarcation of a space labelled the Argentine Antarctic sector was no more than an elaborate administrative fiction. Vast areas of this region have never been seen or walked over. As with their Chilean and British counterparts, Argentine survcyors and naval personnel were instructed to undertake topographic and geological surveys whilst at the same time ensuring that evidence of permanent occupation (bases, plaques and flags) were visibly evident. In Buenos Aires, however, President Perón frequently referred to Argentina's "inalienable rights" in the Antarctic and the South Atlantic.

More broadly, these concerns over asserting Argentine sovereignty over the Antarctic and South Atlantic had to be understood within Perón's vision of Argentina in the post-war period. Argentina was adamant that the

Antarctic and Southern Ocean had to be "decolonized" so that imperial powers such as Britain could be removed from the American continental region. These anti-colonial sentiments may have been opportunistic given Argentine territorial claims and the lack of indigenous populations, but they did contribute to Perón's vision of an independent Argentina. To this end, Perón approved the creation of the Antarctic and Malvinas Islands Division within the Ministry of Foreign Affairs and Worship in 1948 so that Argentine expeditions could be related to a strategy of claim improvement. Related to this last point, Perón had argued that Argentina had to pursue a "Third Way", an approach to foreign relations which avoided being trapped either within the competing orbits of the superpowers or the imperial superstructure of Britain (see Melensky 1978). This was modified in the late 1940s as the military and institutional bases of a Pan-American security system began to materialize, especially after the 1947 Rio Treaty of Inter-American Reciprocal Assistance (Conil Paz and Ferrari 1966:165–6, *American Perspective* 1947). The latter was signed by 21 American Republics, including Argentina, and pledged that an armed attack by any state against an American state should be considered an attack on all of these states. Article I stated that the geographical jurisdiction of the Rio Treaty extended southwards and incorporated the Antarctic continent. Argentina and Chile referred to the provisions of Article IV to argue that Britain's robust defence of the Falklands Islands Dependencies (such as dispatching a warship to the region) in the 1940s and 1950s constituted a threat to an American state. Whilst statements of support were issued from the National Assembly of Venezuela (10 March 1948), the Brazilian newspaper *Correo da Manha* (17 March 1948) criticized Argentina and Chile for raising the issue in a recent pan-American meeting in Bogota. The United States later rejected Argentine claims that the Rio Treaty could be invoked against Britain in Antarctic matters. It was argued by the Truman administration that the Rio Treaty had been intended to forge an anti-communist alliance against the Soviet Union rather than a post-war ally (Black 1986).

A consequence of the American reluctance to invoke the Rio Treaty against the British in the Antarctic was that Argentina and Chile began to forge new contacts over Antarctic matters. Significantly, both states began publicly to articulate the concept of an American Antarctic (Child 1979, 1988): a geographical place which was defined by its proximity to the South American mainland and which should be free from the activities of external powers such as Britain. A joint declaration with Chile released in March 1948 stated that: "Both governments will act in mutual accord in the protection and juridical defence of their rights in South America Antarctica, included between the meridians of 25 degrees and 90 degrees west longitude, in which the territories the Republics of Argentina and Chile recognise their indisputable rights of sovereignty" (cited in Pittman 1981: 681).

Geopolitical and geographical writers such as Isola and Berra produced *Geopolitica de Argentina* which also cited, approvingly, the notion of a South American Antarctic (Isola and Berra 1950:25). However, it should be remembered that Argentine–Chilean co-operation was often fragile in the 1950s in the face of great uncertainty over the role of the superpowers in the region. Both states later produced maps and charts depicting their national claims whilst excluding references to a South American Antarctic. One of the few achievements in Anglo-Argentine–Chilean relations during this period was the signing of a naval agreement in November 1948 which stipulated that warships would not be sent below the latitude 60 degrees South. This was a major diplomatic success because of fears that the Argentine military personnel who accompanied expeditions might clash with British armed forces attached to visiting warships.

In 1951, Argentina created the Instituto Antarctico Argentino (IAA), under the jurisdiction of the Ministry of the Marine. The intention of the IAA was to centralize the planning and co-ordination of the Argentine Antarctic programme. Perón had earlier argued that he: "hoped that in time the justice of the Argentine claim would be generally recognised. If not, Argentina might one day have to use force" (cited in Dodds 1994b: 271). In the same year too a major expedition was dispatched to the Antarctic peninsula with the object of creating a new permanent base called General San Martin. The Argentine press invented the term "*Argentarctica*" whilst reporting the achievements of the expedition. Later it was argued that: "the country needs it [San Martin] in order to continue with scientific research on the western coastline of the Antarctic peninsula" (Borten 1970:7). Predictably, the British newspapers were less sanguine about Argentine Antarctic activities; the *Daily Mail* argued that this was evidence of a "New Perón grab" (1 May 1951). At this point, Anglo-Argentine relations in the South Atlantic and Antarctic were fragmenting as Perón's government began to invest more money and materials into the Antarctic programme (Roucek 1951, Illingworth 1953). The old trading relationship was declining in significance and geopolitical competition was emerging as the dominant feature of Anglo-Argentine connection.

Significantly, Argentine endeavours in the Antarctic were supported by a vigorous government campaign and enjoyed widespread public and media support. Perón's speeches on the Antarctic were numerous after his election in 1951 and usually included claims that the Argentine Antarctic sector was a natural part of the Republic and was currently imperilled by the occupation of the British. Perón refused to submit the disputed claims with Britain to the International Court of Justice at The Hague because: "it reaffirms its possession de facto and de jure over its sector of the Antarctic, and cannot therefore submit its territories which legally belong to it and of which it has taken effective possession, to the arbitration of the above-mentioned international organism" (International Service of Argentine

Publications 1954:50). In a speech to the Argentine Congress in May 1952, Perón outlined his aspirations for the Antarctic: "those who discuss our rights have thousands of years behind, but perhaps very few ahead. We have a century behind, but many in front of us" (cited in Dodds 1994a:272). His comments came in the aftermath of the infamous Hope Bay incident when Argentine and British personnel clashed over the occupation of a base in the Antarctic peninsula. It was later alleged that Argentine soldiers had fired over the heads of British surveyors and scientists (Beck 1986: 35–6). Perón, however, admonished Uruguay for having recognized British claims to the Antarctic in October 1952.

This incident was to raise new fears that the polar continent would become a source of major conflict between the two parties. From the Argentine perspective, it was argued that Britain was contributing to a climate of violence because the British government sought forcibly to remove South American personnel and their bases from the Antarctic. In 1953, the Foreign Office in London instructed HMS *Snipe* to arrest and deport Argentine personnel from Deception Island. The British also dismantled Argentine and Chilean bases in and around the Antarctic peninsula. The Argentine foreign minister Jeronimo Remorino later noted that: "undoubtedly the British aggression in the Antarctic violates the Rio de Janeiro Pact and offends not only Argentina but also the principles of continental defence" (*Ottawa Journal* 26 February 1953). The appeals to the Rio principle for continental security were ignored by the United States, which was intent on establishing a possible international agreement over the future of the Antarctic. But other South American states such as Brazil and Colombia had supported Argentina's request for military support from the Inter-American security system.

Interestingly, the Canadian government, which was informed of the British actions, was uneasy over the British response. The Department of External Affairs in Ottawa confirmed in September 1953 that these evictions might have "very unfortunate repercussions for relations with Latin America" (cited in Dodds 1996c). The repercussions in Argentina were modest, however. The Argentine Antarctic programme continued to construct new bases in the Antarctic peninsula, and in 1955 the base General Belgrano was completed in conjunction with the purchase of a new ice breaker, *General San Martin* (Hanessian 1965:12). A significant development in March 1957 was the further legal incorporation of the Argentine Antarctic when it was announced that a "new" administrative territory had been created, The National Territory of Tierra del Fuego, the Antarctic and the Islands of the South Atlantic, with a capital city based in Ushuaia (*Boletin Official* 19 March 1957). The geographical area of jurisdiction also incorporated the Falklands/Malvinas, the South Orkneys and South Georgia. This preceded further British requests to submit the question of disputed sovereignties to the International Court of Justice. Moreover, the

Argentine government had decided to participate in the forthcoming International Geophysical Year (IGY 1957–8) after creating a National Commission for the Argentine contribution to the IGY. Argentina's involvement in the IGY had been earlier secured through the 1955 Agreement in Paris that these scientific activities would not be interpreted as weakening the rights of claimant states in the Antarctic. In both cases, these moves were driven by the conviction that the Argentine Antarctic sector was indisputably part of the Republic and attempts to weaken their "rights" in the region would be opposed. The debate over a possible extension to the IGY was indicative of Argentine concerns in the Antarctic about the intentions of the Soviet Union in the aftermath of the IGY. Argentina was worried that without a new "Paris Agreement", the Soviets might be tempted to press a formal claim to Antarctica in the absence of either an extended IGY or a binding agreement on territorial rights. Hence the chief Argentine delegate during the IGY negotiations, Admiral Panzarini, reminded his audience at the 4th CSAGI meeting in Paris in June 1957 that Argentina was willing to contemplate an extension to the IGY: "that owing to the proximity of the part of Antarctic facing Argentina and owing to the fact that Argentina has permanent bases in Antarctica, the extension for one year of the IGY activities [ie after December 1958] there, would but require an insignificant additional effort" (cited in Hall 1986).

Between 1957 and 1959, Anglo-Argentine relations had improved according to Argentine and British official assessments (Dodds 1994b, 1996c). The British government considered the new Frondizi government more moderate than the Perón government over the Antarctic and the Falklands/Malvinas. In a major review speech on foreign policy, Foreign Minister Carlos Florit announced in May 1958 that Argentina would continue to press for the protection of Argentine Antarctic sovereignties whilst seeking to promote scientific and administrative co-operation with other interested parties. However, the British Embassy in Buenos Aires was sanguine: "bilateral contacts have been reasonable given always the fact that she [Argentina] has felt it necessary to give us written reminders of her claims to sovereignty over the Falklands and the British Antarctic" (cited in Dodds 1994b:278). In both countries, there was general political agreement that, if the forthcoming negotiations in Washington failed, then the 1948 Naval Agreement would be renewed again in 1959. Argentina approached the negotiations, according to one of their chief Antarctic negotiators Roberto Guyer, with the hope that there would be a legal and political accommodation on the issue of territorial sovereignty, fearing that the parties were on "the threshold of a conflict which could have serious political consequences" (Guyer 1983:270).

By the late 1950s, the Eisenhower administration had committed the US State Department to devising an international agreement for the Antarctic to deal with the threat of either Anglo-Argentine–Chilean rivalry or the possible involvement of the Soviet Union. Argentina consistently opposed any internationalization of Antarctica as well as Soviet involvement in the polar region. For example, the *New York Times* reported that the Argentine Foreign Ministry had issued a rapid note of protest against a Soviet landing in Traverse Islands (near the Antarctic peninsula) which it claimed violated Argentine territorial sovereignty (12 January 1958). The Frondizi government remained adamant that Argentina would resist any attempts to weaken or diminish its territorial claims to the polar continent. In the formal reply to the American offer to discuss the future of the Antarctic, the Argentine government reaffirmed that:

> The Government of the Argentine Republic wishes, above all, to establish clearly that it would not consider attending an international conference in which questions related to Antarctica might be discussed, if this could be interpreted by the participating countries as affecting in any way its inalienable rights of sovereignty over the Argentine Antarctic sector, the Malvinas, the South Georgia Islands, and the South Sandwich Islands. (Canadian Archives, Department of External Affairs, 1958)

The reason for this apparent intransigence needs to be understood within the interrelated contexts of Argentine foreign policy priorities and a mounting concern that the Antarctic would become a site of tension and conflict between the superpowers. Frondizi's pursuit of an autonomous foreign policy which included the promotion of industrial development, political non-alignment and freedom from foreign interference also encapsulated a hope that so-called "UN interference" in the management of the Antarctic would be avoided.

In the run-up to the Antarctic conference in October 1959, the representational geographies of the Antarctic had been fundamentally altered. The territorial intransigence of South American states coupled with fears of a new Cold War had transformed the continent into a place invested with considerable strategic qualities and political values. Argentina's preoccupation with territorial claims was not, therefore, out of kilter with wider international trends. Whilst the International Geophysical Year had demonstrated that nations could co-operate with one another in the fields of science, logistics and technology, it had not resolved the issues of territorial sovereignty or the international control of the Antarctic (Beck 1986, Hall 1986). The debates over the possible extension to the IGY had revealed that Argentine delegations were more concerned that after December 1958 the probable activities of the Soviet Union would have implications for Argentine national sovereignty rather than the promotion of science. The challenge for the Eisenhower administration was, therefore, how to transform these moments of co-operation into a foundation for a new regime in order to prevent international conflict whilst at the same time negotiating with countries which stated their territorial claims.

Argentina and the 1959 Antarctic Treaty

The opening statement by the Argentine representative, Ambassador Scilingo, at the negotiations for an Antarctic Treaty on 10 October 1959 informed the representatives from 11 other states that Argentine claims to the Antarctic were:

> A natural extension of the Argentine Republic, [it] has, for many decades, formed an integral part of its territory . . . Argentine military men and Argentine men of science have performed many deeds of heroism in the Antarctic wilderness and some have given their lives in the unceasing process of their outstanding technical and scientific work . . . The Argentine Republic is the first actual and continuous occupant of the area, which has been incorporated administratively into the Tierra del Fuego government district for some time, and is attending this conference with all these rights. (SPRI Archives 1959)

The Argentine delegation had arrived in Washington with a clear commitment to protecting its territorial claims in the Antarctic and South Atlantic. As with their Chilean and Australian counterparts, the Argentines were adamant that any agreement over the future of the polar continent would not affect their territorial rights (Casellas 1981:66–71). It should also be recalled that Argentine political observers were aware that recent Brazilian geopolitical writings had proposed that the South American Antarctic be divided along the basis of a so-called "Frontage Theory". The legal standing of this proposal was highly dubious, but it did contribute to a hardening of Argentine resolve over their territorial rights (Casellas 1981:33–4). As the *New York Times* stated on 14 October 1959: "The fear that bases there might be used for warlike purposes has made the continent's future of vital concern to neighboring lands. The territorial claims by some nations especially Argentina and Chile are highly emotional issues in domestic politics."

Ironically, in spite of the fears of the British and American delegations over Argentine intransigence regarding territorial claims and Soviet mischief making, it was the French delegation which rejected the compromise over such rights. The proposed Article IV of the Antarctic Treaty had intended simultaneously to acknowledge these territorial claims whilst demanding that these claims would not be improved upon under the jurisdiction of the Antarctic Treaty. It also stipulated that countries such as South Africa which had not pressed a claim to the Antarctic did not have to acknowledge the claims of other nations. The Canadian embassy in Washington reported that: "The French, quite unexpectedly, had arrived in Washington with rigid instructions to oppose the inclusion of Article IV in the Treaty, on the grounds that they were not prepared to agree to anything which might imply derogation of their own sovereignty" (National Archives of Canada, Department of Exernal Affairs 1959). There were also problems over Argentine objections to the freedom of movement within the Antarctic Treaty area because that could be considered tantamount to allowing occupation of her sovereign territory by foreign expeditions (Scilingo 1973, Fraga 1979). The question of sovereign rights was a crucial element in the negotiations leading to the Antarctic Treaty. Argentina and Chile had been adamant that their rights be unaffected by any new international regime. In a confidential memorandum, it was reported that the British foreign secretary, Selwyn Lloyd had had a secret meeting on 16 November 1959 with his Argentine counterpart who had noted:

> The Argentine government was very concerned at the prospect of the Russians' having free access to the Argentine sector of Antarctica. His [i.e. Frondizi] government would like laid down that the Russians could only enter the sector for definite scientific purposes and should withdraw when this had been accomplished. (British Archives, Foreign Office 1959)

The British response to Argentine and Australian fears over Soviet involvement in the Antarctic was later accepted by both parties. The reasoning was surprisingly simple: it was better to bind the Soviet parties to some agreement over scientific activity rather than none at all. At the time of this meeting, India had raised objections to the Washington Conference and Peru had declared that there existed a Peruvian Antarctic sector between 77 degrees and 90 degrees West (Mercado Jarrin 1984). This Peruvian claim overlapped with Argentine, British and Chilean polar claims.

The peaceful use of Antarctica was one of the central objectives of both the preparatory meetings and the Washington Conference. Until the Washington Conference, the 12 interested parties had not agreed that the polar continent should be a zone of peace and therefore demilitarized. This understanding of demilitarization was later expanded to consider the possibilities of nuclear weapons, tests, waste and explosions. Argentina, with the support of other southern hemispheric nations such as South Africa and New Zealand, raised the issue of a nuclear prohibition zone with the proposal for a complete ban on nuclear explosions within the treaty area (Scilingo 1973). Between mid-October and mid-November 1959, the issue of nuclear prohibition generated much debate within the Washington Conference. The Soviet Union was adamantly in favour of the ban, the American and British delegations were against a ban, whilst a small group of nations such as Belgium, Japan and Norway were in favour of a partial ban. The Argentine delegation with the support of Australia and Chile pressed for a complete prohibition and finally, on 28 November 1959, it was agreed that nuclear explosions in the Antarctic would be prohibited (Beck 1990a).

Argentina's determined effort to prohibit nuclear testing and explosions in the Antarctic received widespread support from other southern hemispheric nations. The attitude of the Chilean representative was indicative of that support: "Let's carry on tests in Siberia, in the Pacific, in the Sahara and in Australia if desired, but not in Antarctica. Why do we need to test in

Antarctica?" (cited in Beck 1990c:196). This position was further supported by sections of the media covering the Conference proceedings. As the *Washington Post* acknowledged in October 1959: "Considerations of this sort are bound to arise in such discussions, and Argentina and Chile being closer geographically to the frozen continent have perhaps extra reason to concern themselves with its use" (20 October 1959).

In the aftermath of the Antarctic Treaty negotiations, Argentine reactions over the final outcome were mixed. Initially, a radio broadcast from Radio Splendid (15 December 1959) in Buenos Aires affirmed that Argentina's territorial rights to the Antarctic had been unaffected by the Washington Conference. Roberto Guyer, the deputy leader of the Argentine delegation later noted in July 1961 that:

> The Antarctic Treaty, incidentally, brings out clearly the fact that we are neighbours at the South Pole. There are many affinities between Argentina and Australia . . . Because of its geological continuity, its geographical proximity, and the efforts exerted over there for many years, the Argentine Antarctic is part of our national territory . . . cordiality and harmony have been outstanding characteristics of scientific work in Antarctica. (ATCP 1961)

Later, however, Alberti Candioti, a former Argentine ambassador, argued in the *Buenos Aires Herald* that: "This conference like the one held last year in Canberra (ie in 1961), is part of a programme to internationalise the Antarctic, a plan which should be energetically opposed by all Argentines . . . We have already lost some of our rights in the Antarctic" (19 July 1962). For most Argentine commentators, the Washington Conference had been successful in terms of the demilitarization of the continent and the prohibition of nuclear testing. Indeed, it was these concessions that enabled the Argentine government later to argue that Argentine claims to the Antarctic had not been compromised (cf. Candioti 1960). The New Zealand delegation's report on the Washington Conference confirmed that the Argentine delegation had never pressed their objections over territorial claims to the point of breakdown either because of a fear of upsetting the 11 other parties or a concern over possible UN involvement.

Science, territorial claims and resource interests in the Antarctic and Southern Ocean

The Argentine research programme in the aftermath of the Antarctic Treaty was varied, ranging from participation in the International Year of the Quiet Sun (1964–5) to reinforcing research activities at a number of bases scattered around the Argentine Antarctic sector (Milia 1978, Fraga 1979). In spite of Article IV preventing parties to the Antarctic Treaty from seeking to alter existing territorial claims, successive Argentine governments often funded high-profile events to ensure that the Argentine

Antarctic sector remained embedded within the public consciousness. An Argentine naval flight over the South Pole in January 1962 was widely interpreted in such a manner. Two planes left Ushuia and flew south, stopping at Robertson Island and Ellsworth Base before returning to Argentina. The Canadian embassy in Buenos Aires had noted that the Argentine press "interpreted the flight as a reaffirmation of Argentine rights over Antarctica . . . which they describe as no more than a prolongation of the Andean range which makes up one of the borders of this country" (National Archives of Canada, Department of External Affairs 1962). This interpretation was to remain a persistent refrain in Argentine observations despite the provisions of the Antarctic Treaty. The Argentine research programme has concentrated on activities such as studies of the upper atmosphere and oceanographic research which have distinct implications for the operation of navigation and communication systems. Moreover, the armed forces continue to play a key role in supporting the Argentine scientific programme, calling into question both the quality of the science being undertaken in the Antarctic and the principle of demilitarization (Beck 1992:263–4).

The Argentine delegation to the 1961 ATCP meeting in Buenos Aires set an important precedent for future activities. Whilst the delegation urged the other parties to adopt recommendations aimed at ensuring the exchange of scientific information and the evaluation of living resources in the Southern Ocean, it raised objections to the idea of a permanent secretariat to co-ordinate the business of the ATS. Successive governments have argued that such an organization would compromise Argentina's sovereign authority over the development and resource evaluation of their Antarctic sector. The rationale for this hostility to the principle of a secretariat is based on the assumption that such a body would seek to enforce regulations that might compromise national sovereignty. Such a position was apparent throughout the 1964 meeting and repeated again during the 1966 ATCP meeting in Santiago (Myhre 1986). In a wider context, Argentine attitudes towards the institutional development of the ATS during the 1960s were indicative of a relatively "passive" attitude towards polar management (Beck 1986:158). The role of the ATCP meeting, according to Argentine delegates, was to manage the Antarctic in an *ad hoc* manner rather than to develop elaborate mechanisms and problem-solving capabilities.

This Argentine preoccupation with national sovereignty and legal jurisdiction during the 1960s and 1970s was transparently obvious to other southern hemispheric nations. The Australian embassy in Buenos Aires reported to Canberra in August 1962 that:

> The South American delegations were still sensitive about all matters involving, however remotely, questions of territorial sovereignty, although there is some

> evidence of a growing appreciation among certain officials in Argentina of the spirit of Article IV of the Treaty . . . Argentina and Chile were strongly opposed to anything which contained the germ of a permanent treaty secretariat or which increased the capacity of the United States to dominate the Treaty. (Australian Archives, Department of External Affairs 1962)

The "sovereignty games" in the 1960s and 1970s over sharing communication and meteorological information, coupled with debates over the provision of World Weather Watch Stations, were indicative of Argentine concerns that their sovereignty should not be compromised even indirectly by the recognition of another nation's weather station, located in the Argentine Antarctic sector, by the World Meteorological Office.

Argentine attitudes towards the development of living resource regimes and the Agreed Measures for Antarctic Flora and Fauna have also been characterized by a form of geopolitical reasoning which has stressed territorial sovereignty, resource control and a general reluctance for the ATS to devise measures which might be interpreted as compromising Argentina's rights over their Antarctic sector. This concern became increasingly apparent in the 1970s when northern hemispheric nations such as the Soviet Union and Japan began to extend their fishing activities in the Southern Ocean. Both Chile and Argentina were anxious to restrain these activities in the face of their concerns that the living resources of the Southern Ocean would be depleted by advanced fishing fleets. The Argentine delegations to the ATCP meetings and the BIOMASS programme organized by SCAR played their part in devising guidelines for living resource exploitation in the region. The eventual formalization of CCAMLR in 1980, for example, represented a balance between the claimant and the non-claimant ATCP states, the northern hemispheric fishing nations and outside organizations such as the EEC (Orrego Vicuna 1990). Article IV of the final convention restored the moratorium over territorial claims whilst maintaining a degree of legal and political ambiguity over the status of the high seas and the rights of states to end conservation measures should exploitation become viable.

Argentina's accession to the CCAMLR was dependent upon the formal reassurances of Article IV and the position of the claimant states. This concern was expressed by the Argentine chairman at the 1979 Washington ATCP meeting, Angel Oliveri Lopez:

> It is no secret that in our time the problem of Antarctica is becoming increasingly complex, because together we are entering a stage orientated towards the exploitation of its natural resources . . . it is not inappropriate to point out the special significance which the Republic of Argentina attaches to the preservation of the Antarctic environment . . . Argentina, whose continuing activities in Antarctica extend back several decades before the International Geophysical Year – its first base established 75 years ago – cannot give its sovereign consent to any agreement which does not take this political reality into consideration. (ATCP 1979)

The geopolitical significance of Article IV of the Antarctic Treaty was well appreciated by Argentine policy makers and diplomats (Milia 1978, Fraga 1979, Palermo 1980). Argentine consent for any resource agreements covering the Antarctic and the Southern Ocean were dependent upon the rights of claimant states being formally recognized. The negotiations over resource exploitation and management were important because of the direct implications for the regulation of national maritime space and the rights of claimant states to any resources found within that region (Assef 1980, Pittman 1981).

Argentina's involvement in resource regimes was, however, one element in a wider geopolitical and strategic matrix. Jorge Fraga, as a former director of the Argentine Antarctic Institute, argued that Argentina had to secure "effective occupation" of the Argentine Antarctic Sector and that required the assistance of military personnel (Fraga 1979, 1983). One of the most pressing themes to emerge from this examination is the importance of geographical linkage between the Antarctic, the South Atlantic and islands such as South Georgia (Shackleton 1985). As Robin Edmonds of the British Foreign Office recalled after a meeting with Roberto Guyer in 1962: "For their part, the Argentines took a very deep breath and said, 'OK we've signed this treaty in which we have agreed for 30 years, and maybe beyond to, in quotes, 'freeze our sovereignty'. That is Article Four of the Antarctic Treaty. 'But', they said, 'we are not going to give up, in any way, our claim to the Malvinas' " (cited in Charlton 1989:10). This point was reiterated by an Argentine Antarctic delegate who noted in 1961 that: "The Antarctic Treaty does not affect the legitimate rights of our country. Moreover, the Malvinas Islands are not included in the zone of application" (cited in Genest 1985:27). Regional disputes with Britain over the Falklands/Malvinas and Antarctica, coupled with Argentine–Chilean territorial competition, have ensured Argentine intransigence over "effective occupation" and led in part to demands for regional security pacts in the South Atlantic.

Antarctica and the security of the Southern Ocean

The creation of the Antarctic Treaty (1959–61) coincided with renewed Argentine interest in the defence of the South Atlantic in the face of perceived Soviet maritime expansionism in the 1960s and 1970s (Dodds 1995). Whilst there was general political satisfaction with the ATS, Argentine governments had been anxious to promote regional defence and security co-operation in the South Atlantic. In 1956, for example, Argentina had approached Brazil and Uruguay over the possibility of convening a naval conference to consider co-operation and joint exercises. The *raison d'être* for such a request (which was later declined) was that the Soviet Union and communist forces could imperil the shipping lanes of the South

Atlantic (Hurrell 1983). Anti-communism was emerging as the dominant theme in the ideology of civilian and military regimes in South America fostered and supported by the Inter-American Defence System. Argentine military writers argued that the South Atlantic lacked a defensive arrange ment similar to the North Atlantic Treaty Organization (NATO). Alberto Casellas (1974), an Argentine naval writer, called this part of the Atlantic: *El territorio olvidado.* These proposed regional developments were important because even though the Antarctic had been declared a zone of peace (with Soviet agreement), Argentina maintained that her Antarctic claims were intimately linked to the defence of the South Atlantic.

Argentine geopolitical writings on national security, strategic "chokepoints" and the projection of power in maritime space became increasingly significant in the 1960s and 1970s given the three armed interventions of the military into government (Child 1988). Argentine geopolitical writers such as Fernando Milia, Jorge Fraga and Vicente Palermo have argued that the Antarctic and South Atlantic region is a "geopolitical space which integrates Eastern South America, Western South Africa, and the Antarctic continent and the vast sea that these lands delimit" (Milia 1978:1, but also Casellas 1974, Palermo 1980). Improving relations with Brazil meant that Argentina could become less preoccupied with its major continental rival. There was also a growing perception that substantial resources existed in the Southern Ocean and the Antarctic (Hepple 1992). The Antarctic Treaty had not at that stage dealt with the issue of possible mining and mineral exploitation. Argentina's reluctance to discuss resource issues in the meetings of the ATCP in the 1970s reflected a concern that either Argentine sovereign interests would be adversely affected by any convention or that Argentina would be technologically incapable of exploiting those Antarctic resources in comparison with Britain, the Soviet Union, the USA or Japan. Further, the growing importance of the Argentine navy meant that a more maritime focus was perhaps inevitable. Alberto Casellas recorded that the ideas of Admiral Storni were eagerly re-examined as the navy sought to create a maritime consciousness in Argentina (Casellas 1974:17–19). Argentine governments were increasingly concerned too that the resource-rich South Atlantic was a region vulnerable to the geopolitical designs of the Soviet blue-water navy, especially after the Soviet–Cuban occupation of Angola (Kemp 1977). Finally, the disputes with Britain over the Malvinas and Chile over the Beagle Channel contributed to a renewed geopolitical interest in the Southern Ocean and the Antarctic.

A key figure in this transformation was General Juan Gugliamelli and his Instituto Argentino de Estudios Estrategicos y de las Relaçiones Internaçionales which produced the journal *Estrategia.* Whilst there has been some debate over the intellectual merits of this journal, there can be no doubt that Gugliamelli was a major figure in shaping Argentine polar consciousness as chief of the Escuela Superior de Guerra and then later the head of the Fifth Army Corps (Pittman 1981, Child 1985). Under the Ongania government (1966–71), Gugliamelli was also secretary to the Consejo Naçional de Desarollo (National Council of Development). Whilst it is difficult to trace the influence of specific ideas into government policies, Gugliamelli's writings undoubtedly reflected the prevailing mood in military-dominated Argentina. His nationalist writings either on countering Brazilian expansionism or promoting regional development in the south reflected the view that Argentina was "peninsular" and composed of three parts: the continental northern interior, the Atlantic ocean extension (Mar Argentino) and the Patagonian and Antarctic peninsulas (Gugliamelli 1976:5–19). In order to protect the territorial integrity of Argentina, Gugliamelli argued that the military would have to counter Brazilian expansionism in the South Atlantic and in the La Plata basin as well as preventing Chile and Britain from developing their claims in the Falklands/Malvinas, the Beagle Channel and the Antarctic (Cirigliano 1975).

An integral part of Gugliamelli's programme for national action was the representational transformation of the South Atlantic from a geographical backwater into a strategic priority. In the aftermath of the Antarctic Treaty, Argentine writers began to discuss the factors that were leading to the changing significance of *Atlantartida*. There are four major strategic arguments which played a key role in this transformation (Dodds 1994a). The first can be described as the Cape of Good Hope argument. In conjunction with South African geopolitical writers, Argentine military officers argued that the shipping lanes around South Africa were threatened by the Soviet blue-water navy. Supertanker traffic, with oil from the Middle East, could be disrupted or, even worse, destroyed by that naval force. It was vital, therefore, that in the absence of Western naval support countries like South Africa and Argentina combined to defend these shipping lanes. Second, the Panama Canal Scenario, based on the assumption that the Canal might be blocked by terrorists. In that event shipping would be forced to travel around Cape Horn in order to cross between the Atlantic and Pacific Oceans. The strategic importance of Drake's Passage would be transformed and the Falklands/Malvinas would play an essential role in the maritime defence (Cosentino 1970). The perceived resource value of the Antarctic and the Southern Ocean was also used to argue that Argentina had to maintain an active scientific interest. General Osiris Villegas on a flight over the Antarctic in November 1967 made this point: "My spirit as an Argentine and as a soldier is filled with emotion when I contemplate this part of our land which is so distant and yet so much our own . . . defending the southern frontiers of our country with our navy and civil scientists" (cited in *Correo del Pinguino* 1970 March:5). Fourth, it was considered imperative that Argentina had to be an anti-communist and a regional hegemon in the South Atlantic in order to preserve the South Atlantic and the Antarctic from communist "infection".

Given these views, the Argentine approach to the security of the South Atlantic and the Antarctic was based on the assumption of maintaining a substantial scientific Antarctic programme. In spite of Argentine fears to the contrary, the ATS has successfully maintained the polar continent as a demilitarized space. In terms of their territorial claims, Argentina has persisted with its demands that Article IV of the Antarctic Treaty does not alter its claims to sovereignty. The creation of the Direccion Naçional del Antarctico, under the Ministry of Defence, in 1969 reflected the determination to protect those polar claims. A former director, Admiral Jorge Fraga, in his book *Introduccion a la Geopolitica Antartida*, firmly restated Argentina's claims in contrast to the "weak" British and Chilean sectorial claims. The Argentine research programmes in earth sciences, biology and meteorology were reviewed for the purpose of demonstrating Argentine activity in the region. Argentina also ensured that its agreement to the Convention for the Conservation of Antarctic Marine Living Resources in September 1980 was accompanied with a formal note declaring that its rights to sovereignty were not affected by the CCAMLR zone of application (Genest 1985).

Under the Proceso military regime (1976–82), Argentine interests in the Antarctic and South Atlantic were undoubtedly at their most militant. Within the Antarctic Treaty System and the formal consultative party meetings, the Argentine delegations consistently reminded their audiences of the pertinence of Argentine territorial sovereignty. Argentine expeditions, for example, despatched pregnant women in 1978 to their bases in order to declare that the subsequent newly born babies were indeed Argentine citizens (Pittman 1981, Child 1988). The Ministry of Education introduced a compulsory course on geopolitics in secondary schools in 1979 which included teaching on "Argentina y el Atlantico Austral: Soberania Naçional Argentina, Prioridad Naçional" (cited in Pittman 1981:865). According to the document Circula General Number 48, teachers were instructed to enhance the geopolitical consciousness of school children thus improving awareness of Argentina's claims to the Antarctic and the South Atlantic. Geopolitical writers such as Alberto Assef in *Proyeccion Continental de le Argentina* concluded that:

> The decisive recovery of the Malvinas will nurture the spirit of Argentines. It will be an injection of will and a stimulus for our vocation towards our destiny. It will provoke and move the Nation, making a new era of honour and progress. It will cement our domination of maritime spaces, it will affirm our rights on the white continent, it will forge the physical integrity of the country, and it will, above all, elevate the spirit of the Nation, giving it faith in itself. (Assef 1980:232)

These assertive actions complemented the renewed concerns of the military to promote a regional pact in the South Atlantic. In the aftermath of Argentine involvement in the South Atlantic Maritime Area Coordination (which ensured that naval data was shared with Brazil and Uruguay) in 1966, their attempts to construct a South Atlantic Treaty

Organization (SATO) in the 1970s reflected the high geopolitical and strategic appraisals given to this region (Fraga 1979:38). In 1976, the Argentine government raised the issue of a SATO with other South American states and South Africa (Coker 1982, Bontori 1988). Earlier naval and diplomatic co-operation in 1966 and 1968 was increased after Portugal had left Mozambique and Angola in 1975. The large-scale injection of Cuban troops into Angola gave rise to new fears in Buenos Aires and Pretoria that the South Atlantic (as well as South Africa) was threatened with Soviet-influenced forces (Sidaway and Simon 1993). The Argentine newspaper *La Naçion* noted after a naval conference held in Buenos Aires in September 1976 that: "Our relationship with South Africa must be reconsidered. South Africa is a bastion against communist infiltration already present in the South Atlantic . . . There are only three countries, who by their culture and their tradition are part of the western world, have a geographical situation which enables them to play an important role in the control and the protection of the southern Atlantic: Argentina, Brazil and South Africa". Other Argentine newspapers such as *Hoy* carried details of a so-called South Atlantic "pact", whilst the Brazilian and American governments always denied that a naval pact had ever materialized (see Hurrell 1983).

The idea of a naval alliance returned again when Argentina hosted a conference in Buenos Aires in May 1981. The academic and military participants included most of the then pariah states: Argentina, Chile, Paraguay, Israel, Taiwan and South Africa. A naval pact, especially with South Africa, was strongly supported amongst the Argentine military at the time when Argentina was embroiled in conflict with Chile over the Beagle Channel and later with Britain in the Falklands/Malvinas. Whilst these subsequent hostilities did not disrupt the proceedings of the ATS, it did lead to fears that conflict might spread to the Antarctic and the southern part of the South Atlantic (Joyner 1985). General Galtieri, the then head of the Argentine junta, was reported as having argued that the recapture of the Falklands/Malvinas was "merely the beginning of the reaffirmation of Argentina's right to assert territories" (Child 1985:61). Concern was felt by British policy makers and scientists involved with the Antarctic given that in the 1981–2 season, 132 military personnel were reported to be stationed on Argentine polar bases compared to 55 civilian scientists (cited in Beck 1986:71). The subsequent collapse of the Argentine military regime after the Falklands War in 1983 helped in part to allay those fears.

CRAMRA, common heritage and the Protocol

The immediate political context for the Alfonsin government was the defeat of the Argentine forces in the Falklands/Malvinas and the human rights abuses of the Proceso regime (Russell 1988, Marcella 1990). In terms of the former, the Argentine geopolitical literature and the print media were preoccupied with the strategic and symbolic consequences of defeat (Reboratti 1983). An interesting dimension to these discussions had been the implications for the Antarctic and Argentine claims (Child 1985, Joyner 1985, De Hoyos 1988). The argument that control of the Falklands/Malvinas was critical to protecting and ensuring the shipping lanes of the South West Atlantic and the Argentine Antarctic sector were further strengthened by the belief that the United Kingdom's claim to a British Antarctic Territory would weaken if denied ownership of the Falklands (Beck 1988). The planned withdrawal of the ice patrol ship HMS *Endurance* in 1981 was considered disastrous by experienced commentators such as Lord Shackleton who was also swift to recognize these wider regional connections. Mrs Thatcher's decision to increase funding to the British Antarctic Survey in 1983 confirmed Argentine suspicions that the British were intent on raising their profile in the South Atlantic and the Antarctic. Indeed, there were widespread allegations within Argentina that this was the preview to a heightened NATO interest in the South Atlantic coupled with uncertainty over polar mineral resources (Beck 1988, De Hoyos 1988).

One of the effects of the 1982 Falklands War had been to harden Argentine attitudes towards the protection of their territorial and resource claims. At the same time, considerable faith had been placed on the ATS to manage the Antarctic in the interests of Argentina and other claimant states. This was most clearly evident if one considers Argentina's participation in the mineral negotiations between 1982 and 1988 and the subsequent reaction to a number of Third World states intent on declaring Antarctica a "common heritage of mankind". Whilst Argentina had often sought support from the Third World for the claim to the Islas Malvinas, there was undoubted hostility to these "internationalizing" approaches (as there had been to earlier American and Indian proposals in 1948 and 1956 respectively). In their submission to the UN secretary-general's report on the "Question of Antarctica", the Argentine government argued that:

> it must be stressed that the case of Antarctica contrasts with that of other regions or areas of the planet which have been put forward – mistakenly – by way of comparison, such as the sea-bed beyond national jurisdiction, as well as that of outer space. That is because in the case of Antarctica, there is not, as there was elsewhere, a legal vacuum; here we find a regime consisting of the element outlined in the previous section. (UNGA 1984)

Argentine writers have also argued that the Malaysian proposals for change to the ATS were unacceptable because they would compromise Argentine territorial claims and the international political stability of the Antarctic. In a statement to the United Nations General Assembly in 1982, for example, the Argentine delegation put forward that:

> any comprehensive revision or replacement of the Treaty system may destroy it, to the detriment of international law and order, and could have grave consequences for international peace, security and co-operation. It would be somewhat unrealistic to think that, in the present world situation, a new or better legal regime could be agreed upon for Antarctica. Undermining the Treaty could lead to an arms race in the region and to new territorial claims, with resulting conflict. (UNGA 1982)

Argentine hostility to the elements of internationalization within the ATS and even the provisions under Article IV is well documented as the official records of the 1959 Antarctic Treaty illustrated earlier. Later, the Argentine delegation played a key role in ensuring that the 1976 Non-Aligned Conference held in Colombo did not include the Antarctic in its final proceedings concerning the regulation and management of global resources (Beck 1986:277).

The rationale for Argentine enmity to the common heritage proposal was further fuelled by suspicions that other states wished to encourage either a broader membership of the ATS or United Nations administration of the Antarctic region. This was keenly felt in 1983 when India and China acceded to Consultative Party status. The negotiations over a minerals regime were placing new strains on the ATS, especially given the conflicting interests of claimant and non-claimant ATCPs. Carlos Blanco, head of the Argentine delegation to the 1981 ATCP meeting in Buenos Aires reiterated this concern through the employment of long-standing geopolitical reasoning:

> I must once again reaffirm that Argentina exercises sovereignty in Antarctica. Its rights are well rooted on a permanent occupation, which goes back 77 years . . . In this role as a territorially-minded country, interested in Antarctica, Argentina is today imbued with the crucial importance of the question that concerns us here: the possible exploration for and exploitation of Antarctic mineral resources. (ATCP 1981)

The search for an internal accommodation between the claimant and non-claimant states preoccupied the multilateral negotiations between 1982 and 1988. The Argentine delegation had been scathing about earlier attempts to draft a proposed regime for mineral regulation at the 1979 ATCP meeting in Washington (ATCP 1979). The omission of the "continental shelf" within the draft texts was considered to be a fundamental weakness of the possible zone of application of a minerals regime. Moreover, the Argentine delegation at the 1981 and 1983 ATCP meetings argued that the rights of claimant states had to be acknowledged through the construction of a regime which allowed a degree of internal accommodation and mutual recognition. The 1983 ATCP meeting in Canberra also witnessed the Argentine delegation raising further sovereignty-related issues: the legal status of Antarctic ice and the labelling of their weather stations as "Antarctic meteorological centres" in relation to the World Metorological Organization's World Weather Watch programme (ATCP 1983).

At subsequent ATCP meetings, the Argentine delegations were again preoccupied with the protection of their national rights in the Antarctic in view of the mineral negotiations under the chairmanship of Chris Beeby. In contrast to other southern hemispheric nations such as Chile and Australia, the geopolitical reasoning underlying their public positions within the ATS has been territorially militant and, less obviously, preoccupied with the environmental protection of the Antarctic. The Argentine Ministry of Defence, however, was instrumental in the creation of a political programme during 1985 for the expressed purpose of raising the profile of Argentina's scientific and political contribution to the Antarctic region (Direccion Naçional del Antarctico 1988). In part this was related to South American concerns that the ATS was not responding effectively to the challenges posed by resource management and the debates within the UN General Assembly. This theme was reiterated by the Argentine delegation to the 1985 and 1987 ATCP meetings in Brussels and Rio de Janeiro respectively. It also coincided with Chile's demands that the ATS should respond more efficiently to the criticisms of other states and NGOs as well as release information to interested parties.

Argentina's position throughout the CRAMRA negotiations had been supportive of the Beeby proposals to construct a new regime which would safeguard *inter alia* the territorial claims of claimant states. There is no question that the Argentine delegations to these meetings favoured a convention which would have allowed closely regulated exploration and development of mineral resources in and around the polar continent. When the convention was open for signature in November 1988, most of the claimant states were satisfied that their sovereign rights had been sufficiently recognized. An Argentine declaration contained within an Anglo-Argentine–Chilean document dated 17 March 1989 reiterated these common concerns:

> In signing this Convention [CRAMRA], the Argentine Republic reaffirms its sovereignty over the Argentine Antarctic sector and its related sovereign rights and coastal state jurisdiction, which sovereignty, rights and jurisdiction are safeguarded by Article IV of the Antarctic Treaty and Article 9 of this Convention. (cited in Beck 1994c:17)

However, a string of ship-related accidents and changing public opinion in countries like Australia and France caused an erosion of international support for CRAMRA. The sinking of the Argentine ship *Bahia Paraiso* off Anvers Island in 1989 coincided with the *Exxon Valdez* running aground in Alaska and discharging millions of gallons of crude oil into the surrounding seas and coastlines. Once Australia and France had announced their formal objections to CRAMRA in August 1989, support for a new convention on environmental protection began to materialize amongst a number of European ATCPs such as Italy and Sweden.

The controversies surrounding the rejection of CRAMRA and the negotiation of the Protocol between 1989 and 1991 are covered in some detail within the subsequent Australian and New Zealand chapters. The decision to reject the CRAMRA document was taken reluctantly by the Argentine government, which had been anxious to secure agreement on minerals and resource rights for claimant states. The crucial meeting at Vina del Mar during November and December 1990 witnessed the Argentine delegation linked in an alliance with other nations strongly supportive of CRAMRA: Norway, the United Kingdom, the United States and Uruguay. During the meetings, a so-called "five-power" draft protocol was submitted by these nations which contained details of provisions and measures dealing with environmental pollution and environmental impact. The fundamental difference between that protocol and those submitted by the anti-CRAMRA group was that the former were completely opposed to a total ban on mining (Kyriak 1996:329–30). As a consequence of the sterling endeavours of the Norwegian negotiator Rolf Trolle Andersen, a compromise settlement was reached with the opposing camps (Andersen 1991). The Argentines later agreed to a series of measures designed to improve the environmental protection of the Antarctic and the institutional framework of the ATS in the face of fears that growing criticism from ecological movements and other states might undermine the ATS. The eventual adoption of the Protocol in 1991 was to a large part due to the fear that no agreement would be reached. As with their Chilean counterparts, practical political concerns carried more weight than a heart-felt commitment to the environmental protection of the Antarctic. As Juan Carlos Beltramino recalled: "A degree of credit for the adoption of the Protocol must be given to the non-governmental organisations which were committed to the permanent environmental protection and conservation measures through their reports to Antarctic Treaty Consultative Meetings and lobbying of individual Consultative Parties" (Beltramino 1995:105).

The formal adoption of the Protocol in 1991 did not mean that the issue of minerals and resource rights had diminished in political saliency. The Argentine government remains convinced that the Protocol did not settle the problem of accommodating the competing rights of claimant and non-claimant states (Beltramino 1995). Moreover, Argentine delegations have pressed for two issues to be resolved in the near future: the creation of a permanent secretariat and the effective regulation of polar tourism. On the question of the secretariat, the Argentine government maintains that such an innovation is important if the business of the ATS is to be carried out more effectively. In the light of the pressures faced by the ATS in the 1980s, Argentine delegations to the ATCPs also suggested that the secretariat would enable the Treaty parties to maximize effectiveness if meetings were held in one place rather than rotating on an annual basis (Chaturvedi 1994). It had been proposed that the secretariat should be located in Buenos Aires rather than a northern hemispheric capital city such as London or Washington DC. However, the question of the secretariat has not been resolved, and some countries such as Britain would not accept the proposed Argentine location. The regulation of polar tourism remains a sensitive issue for Argentina because it touches upon national sovereignty and legal jurisdiction. As with their Chilean counterparts, the Argentine delegation at the ATS proposed that tourism should not only be regulated under a separate annexe to the Protocol but that questions of liability should also be clarified in the light of the growing number of tourist ships operating from Argentina and Chile (ATCP 1991, 1992).

The Argentine government formally ratified the Protocol in October 1993, but questions concerning the Protocol's capacity to deal with the issue of liability for environmental damage remain a pressing concern for claimant states including Argentina (Beltramino 1995). As Paltridge and Green (1995:4) note:

> The claimant states would not be happy to hand over that right [ie to adjudicate on the matter] to some international body for fear of downgrading their claim to sovereignty. By the same token, the rest of the Treaty nations would not be happy to hand over that right to the claimant states for fear of tacitly recognising the national claims. The liability question is so vexed that the most likely outcome is continual deferral of the matter.

In conjunction with the ongoing controversies over the Law of the Sea and the determination of Exclusive Economic Zones, Argentine governments of the future are unlikely to adopt new forms of geopolitical reasoning. Issues pertaining to territorial sovereignty and resource rights over the Argentine Antarctic sector will be ongoing.

A powerful illustration of these enduring concerns over resource rights came during the negotiations over fishing stocks and marine protection in the Southern Ocean. The selling of fishing licences by the British government in and around the waters of South Georgia had created considerable tension at the meetings of the Convention for the Conservation of Antarctic Marine Living Resources. Whilst Britain and Argentina continue to dispute sovereignty over this island and the Falklands/Malvinas, the British collection of revenue from these fishing activities is resented. The New Zealand delegation's official report in the aftermath of the 1995 meeting of CCAMLR in Hobart also cited this outstanding tension (New Zealand 1995). Unsurprisingly, Argentine delegations have recently stressed the importance of regional resource co-operation with neighbours such as South Africa, Namibia (another state which has experienced problems over fishing rights) and Brazil in forums such as the South Atlantic Zone of Peace and Co-operation. In their 1996 South Atlantic Zone of Peace and Co-operation (SAZPC) ministerial meeting in South Africa, the Argentine delegation reiterated their belief that Antarctica and the Southern Ocean had to be considered part of a wider geopolitical and resource equation

(Mills 1996, Venter 1996). Whilst there is no suggestion that the Argentines would be prepared to launch an invasion to recover disputed regions in the near future, the practical geopolitical reasoning of the Menem government does stress the importance of regaining these "stolen" territories by the end of the present century.

Summary

Anthony Leysens noted that:

> Argentina has always been aware of maritime strategy because of its geographical location, which also forces it to conduct most of its international trade by sea. Argentina's aspirations towards bolstering its position in the region have also been concentrated on extending its influence within the southern South Atlantic . . . Argentinian strategy in the region is based on its goals of controlling the major strategic points in the area to safeguard its claims in Antarctica and the South Atlantic Ocean. These geopolitical premises are specified by the Argentinian navy. (Leysens 1990:29–30)

Argentine political thinking towards the Antarctic and the Southern Ocean has been empowered by public education and the mass media. Geographical ideas about the state and territory are introduced in school textbooks and teachers have to ensure that children are aware of Argentina's three components: mainland, Antarctic and insular. These forms of "patriotic education" have been supported and legitimated by Argentine geopolitical thinking which has stressed the need for new territorial projects in the southern parts of the Republic. President Alfonsin, for example, raised the possibility in 1986 of moving the capital of Argentina from Buenos Aires to the Patagonian town of Viedma. The rationale for such a move (later resisted by civil servants) rested on the belief that a new southern project would help to restore Argentina's place in the world. In that vein, Argentine public culture remains dominated by an intense territorial nationalism whether contemplating the Andean boundary between Argentina or Chile and or claims to the Malvinas. The belief that Argentina has suffered "territorial losses" at the hands of continental neighbours and extra-regional powers remains profoundly important, even though Argentina increased her territorial jurisdiction in the 19th and 20th centuries (Escudé 1988).

Argentina remains an important contributor to the international politics of the Antarctic and one of the two South American claimant nations. A fundamental tenet of Argentine geopolitical and political thinking resides in the belief that the Argentine Antarctic sector is a fundamental part of the Republic (see Map 3). In the absence of interested environmental movements, Antarctic science has often been used in an instrumental fashion to promote these designs rather than foster enhanced understanding of

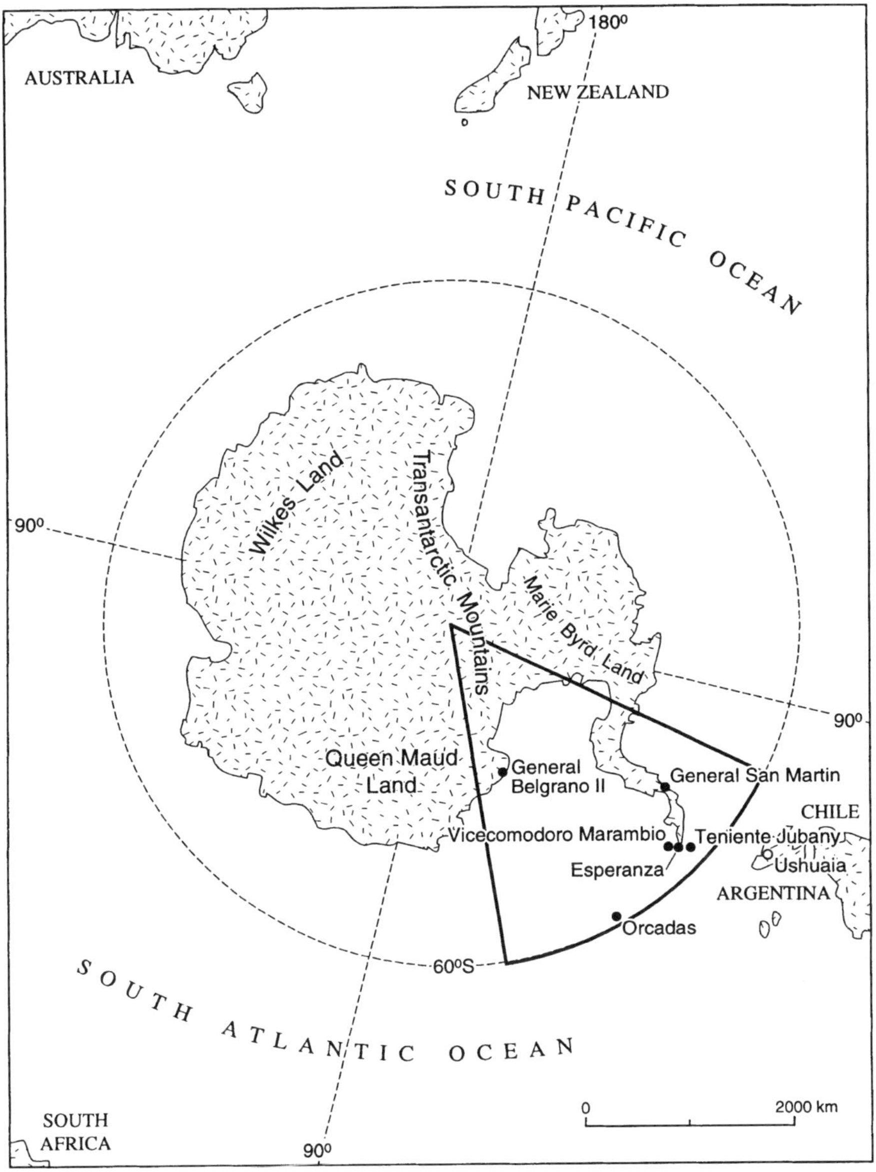

Map 3 Antarctic territory claimed by Argentina, and Argentina's permanent bases on the continent

the region. Whilst the Argentine Antarctic Institute conducts important work on possible ice cap melting, legal arguments over the Argentine Antarctic sector often convey the sense that the consolidation of territorial rights is a top priority rather than environmental protection. Christopher

Joyner's scepticism over the merits of such arguments is succinct and well expressed: "In Antarctic circumstances, being nearest is far from being near. International law, accordingly, is even more cautious when considering the claims issues" (1992:61). The provisions of the Antarctic Treaty System, especially Article IV, have not affected that profound belief in the importance of propinquity and the Argentine Antarctic sector.

4

Australia and the Australian Antarctic Territory

The Australian involvement in the Antarctic has been substantial in terms of polar expeditions, territorial acquisition and scientific endeavour. The creation of an Australian Antarctic Territory (AAT) in 1933 heralded the largest territorial claim to the polar continent. The Order in Council stated that: "part of His Majesty's dominions in the Antarctic seas which comprises all the islands and territories other than Adélie Land which are situated south of 60 degrees South and lying between 160 and 45 degrees East" would be considered to be held by the Commonwealth of Australia (cited in Bush 1982). The intimation of the AAT was itself a product of the changing international politics of the Antarctic and evolving phases of polar co-operation within the British Commonwealth. As with other southern hemispheric nations such as Argentina and Chile, the assumption of a relative geographical proximity to the polar continent has exercised a powerful hold on Australian governments and public opinion alike. As Herr and Davis have concluded:

> As would be expected of a middle power, Australia's foreign policy has tended to define the country's place globally in terms of its capacity to influence regional events of importance to itself . . . Antarctica is undoubtedly a region of significance for Australians. However, a key difference with other regional interests is that Antarctica has only been occasionally an area of dispute in the public consciousness over the last half a century. (1996: 282)

Underlying this sense of continuity has been the conviction, as expressed by successive Australian governments, that their territorial rights to the Antarctic are an integral part of the Australian nation (Triggs 1986, Davis 1992b, Australian Antarctic Division 1995, Herr and Davis 1996, Morgan 1996).

Australia's historical and political experiences of the Antarctic and surrounding oceans have influenced domestic policy making on a par with the two South American states considered here. Whilst regional trading interests in the South Pacific or the Asia Pacific region may be more important in the 1990s, Australian foreign policy has devoted considerable time and resources to the polar continent regardless of whether the government was Labour or the Liberal/National coalition (Cumpston 1995). The controversies over CRAMRA in 1988–9 illustrated the impact of public opinion in shaping government polar policies, and whilst Antarctic policy making remains a restricted enterprise (in the sense of policy involvement and participation) in the SORS, Australian public opinion nevertheless played a major role in ensuring that the Australian government refused to sign that Convention (Elliot 1994a). As Anthony Bergin concluded: "The fact that environmental groups were running a strong anti-mining campaign and appeared to have some bureaucratic allies placed the no-mining position as a real policy option before government. Significantly, there were no powerful groups in favour of CRAMRA . . . As the environmental campaign built up a head of steam the bureaucrats were left behind in political terms" (Bergin 1991:233). Australia has been a major player in the ATS, and it could be argued that the Australian delegations have been the most effective of all the southern hemispheric participants in terms of shaping the conventions and agreed measures. This effectiveness, in part, revolves around Australia's membership of informal political networks such as the Correspondents' Group (which included the USA, New Zealand and the UK) but also reflects the size and personal character of their delegations.

This chapter is divided into six major parts, beginning with an appraisal of the claim to an Australian Antarctic Territory and the ramifications of the legal, political and geographical foundations. Anglo-Australian diplomatic relations are considered as they provide the context for the immediate post-war revival of interest in the Antarctic. Australia's role in the 1959 Antarctic Treaty are examined as are the arguments advanced by her delegates that territorial claims could not be over-ridden by a new international regime. This did not, however, undermine the firm support given for the subsequent demilitarization and denuclearization of the continent. In the aftermath of the Treaty, successive post-war governments considered Antarctic activity part and parcel of a South Pacific and Southern Ocean regional power strategy. The importance of the resource issues in the 1960s and 1970s are evaluated in some detail, as these touched upon the wider regional dimensions. Finally, Australia's role in formulating a new Protocol for the Antarctic in the 1990s and the possible geopolitical significances for such a development are appraised.

Australia and the AAT: imperial inheritance

The first few decades of the present century saw a steady growth in nationally funded expeditions as many states, including Britain, Norway,

France and Belgium contemplated their positions *vis à vis* Antarctica (Beck 1986, Chaturvedi 1996). The Letters Patent in 1908 and 1917 for the Falkland Islands Dependencies illustrated concerns that whaling and sealing (and the revenues derived from these activities) should be subject to British authority. The Australian government, however, supported and dispatched an expedition headed by Sir Douglas Mawson to the Adélie Land and King George Land where he was instructed to implant the Australian flag on the polar ice (Swan 1961). The rationale for the 1911–14 Australasian Expedition was described by Mawson: "It is our special desire to accomplish a complete geographic and magnetic survey between Cape Adare and Gaussberg, a distance of 2,000 miles . . . This portion of the Antarctic continent, to which our efforts are to be directed, is regarded as by far the most important portion of the continent yet to be explored" (Mawson 1911:616). The three-base plan envisaged by Mawson was later compromised by the inclement weather and shifting sea ice. However, the creation of the main base in Cape Denison and a small outpost on Macquarie Island laid the foundations for further Australian Antarctic endeavours. After the 1911–14 expedition, the outbreak of the First World War suspended further expedition planning (Swan 1961, Ferguson 1995). At that time, Australian polar activities were not considered part of a wider imperial design.

In the aftermath of the First World War, successive British governments encouraged Antarctic policies dedicated to consolidating claims to the Falkland Islands Dependencies, the Ross Dependency (1923) and later the Australian Antarctic Territory whilst simultaneously acknowledging the competing rights of France and Norway to other parts of the Antarctic. Throughout these delicate negotiations, which stretched over two decades, those Australian territorial claims to the Antarctic were considered by the British to depend initially on the French and Norwegian reactions to British imperial policy and then later on the activities of the United States in the 1930s and 1940s. The Australian governments of Hughes, Lyons and Bruce were dependent on the United Kingdom because, before the Second World War, all Australian overseas diplomatic representation was limited to London (Cumpston 1995:3). Britain dominated, therefore, the management of imperial affairs, and the Hughes government was intent on influencing imperial policies rather than developing an independent foreign policy.

The background to these delicate negotiations over imperial polar jurisdiction can be traced to a renowned memorandum letter issued by the Secretary of the State to the Colonies Leo Amery to the governor-generals of Australia and New Zealand. Dated 6 February 1920, it stated that:

> HMG have under consideration the future policy of the Empire in the Antarctic regions, and in particular the question of the control of the mainland and of the

> neighbouring waters . . . This area is of great size though as yet of little economic importance except in the case of the Falkland Islands . . . During the war certain of these waters assumed importance owing to the need for glycerine, in which whale oil is rich . . . The increasing radius of action of submarines and aircraft will tend to interest your governments in any territories where preparation might be secretly made for raiding operations . . . with the exception of Chile and Argentina and a few barren islands belonging to France, every inhabited land in the direction of the Antarctic regions is already British . . . HMG have, therefore, come to the conclusions that it is desirable that the whole of the Antarctic should ultimately be included within the British Empire, and that, while the time has not yet arrived that a claim to all the continental territories should be put forward publicly, a definite and consistent policy should be followed of extending and asserting British control with the object of ultimately making it complete. (British Archives, Colonial Office 1920)

This document effectively codified changing Commonwealth perceptions of the Antarctic from an empty isotropic space into a place filled with resource potential and strategic value. The whaling and sealing activities combined with episodes of exploration had been instrumental in this transformation (Beck 1984, Dey-Nuttall 1994). Moreover, the perceptual impact of the First World War had meant that previously remote and inaccessible regions such as the Antarctic could be thought of as part of a global mosaic of competitive economic advantage and strategic significance. The relative proximity of Australia and New Zealand to the polar continent, therefore, was swiftly appreciated by British officials in London. Amery's initial proposal, however, was decidedly in favour of developing New Zealand's polar responsibilities rather than Australia's because the so-called Ross Sea coast was considered the most valuable part of the Antarctic, after the Antarctic peninsula, in terms of resource potential. This part (of the Antarctic) was not only closest to the New Zealand mainland but had also witnessed previous episodes of British whaling and exploration. The fear that France might object to any Australian claim to the Antarctic, given its past involvement in exploration and discovery in Adélie Land, was another reason for favouring New Zealand.

The initial discussions over the possibility of Commonwealth claims and British polar policy were held in 1921 between the Colonial Office and representatives of Australia and New Zealand. It was decided at this meeting that New Zealand would acquire a sector west of meridian 160 degrees West and that Australia should eventually acquire territories to the east. The British issued an Order in Council in 1923 declaring control of a Ross Dependency under New Zealand administration. Over the next five years, the British were engaged in detailed negotiations with the Norwegians and French over their claims to the Antarctic and surrounding seas. The consequences for Australia were largely related to the fact that British officials were not prepared to countenance further claims until the Ross Dependency was recognized by other European states. Later British concessions over Bouvet Island in 1928 and the establishment of a French claim to Adélie Land in 1924 created the legal and political possibilities for an Australian Antarctic Territory.

The Polar Committee created in 1926 was based in London and designed to co-ordinate British and Commonwealth policy and expeditions to the Antarctic continent. Australia and the other interested Commonwealth nations sent representatives to attend these meetings. In conjunction with the Imperial Conferences of 1926, 1930 and 1937, the major issues affecting British and Commonwealth interests such as the defence of trade routes were discussed and decided upon in these forums. As Prime Minister Bruce noted to the Australian Parliament in 1926, "My understanding of the British Commonwealth of Nations is that it is governed by an unwritten treaty of mutual guarantee and therein lies its strength. The different parts of the Empire mutually guarantee one another" (cited in Cumpston 1995:11). During the 1926 Imperial Conference, it was agreed that the priorities for the Commonwealth were (1) to consolidate title to the FID, (2) the Ross Dependency and (3) that Enderby Land (defined as 40 to 52 degrees East) should be brought under the "control of the Commonwealth of Australia if that government is willing". The Australian Liaison Officer, R.G. Casey, urged the British government to annex the Antarctic continent and take possession of a number of Pacific islands close to Australia. Casey later encouraged the Foreign Office to ensure that "the South African firm [Irwin and Johnson] will be given the hint to do some pretty extensive flag planting". The Australian government's willingness to promote polar claims was later to lead Sir Robert Lindsay of the Foreign Office, in an interesting contrast to Amery's vision for an imperial Antarctic, to observe in August 1928:

> We are embarking on a kind of scramble for territory similar to that which took place in the eighties [1880s] for Africa. This is a reversion to obsolete practices which is very much to be deprecated, and started, I should imagine because the two main protagonists, Australia and Norway, are new states who lack the experience we enjoy. I should have thought that there is room for everybody in the Antarctic, unless Australia is deliberately trying to keep everybody out of it except herself, and this is a *reductio ad absurdum*. (Public Record Office, Foreign Office 1928)

In spite of Sir Robert's protestations to the contrary, the Commonwealth Polar Committee was intent on securing British territorial and resource rights in the Antarctic and the surrounding seas. In a letter from Lord Cushendun to the British representative in Oslo dated 18 October 1928, the discussion with Norway over ownership of Bouvet Island had led to negotiations over the future of Antarctic territorial claims. The British had reminded the Norwegians that:

> The discussion of the whole Antarctic question . . . involved two aspects, the first was political and territorial. Considerations of geography alone, with the British

> Dominions of South Africa, Australia and New Zealand projecting down towards the Antarctic continent, gave the British Empire a very special interest in the political rights of those areas, particularly in view of the growing dispositions of the Dominions to interest themselves in the disposal of Antarctic territories . . . the preference of the British government and even more of the Dominions was that the Empire should have no neighbours at all in the Antarctic or in its adjacent islands. (cited in Dodds 1996a:32)

These claims, whether by individuals such as Lord Cushendun or Sir Robery Lindsay to geographical proximity, were, ironically, later to be rejected when the British dealt with similar Argentine and Chilean arguments relating to the Antarctic peninsula. The Argentine declaration of sovereignty over the South Orkneys in 1927 had persuaded the British government to delay the creation of an Australian polar claim on the basis of geographical proximity. Once that issue had disappeared from the diplomatic agenda then the British resurrected these plans. In the early 1930s, with only possible opposition to face from northern hemispheric nations such as Norway and the USA, geographical and cartographic claims to propinquity were again considered valid even though the sector principle had to be modified (Chaturvedi 1996:71–2). Within the Commonwealth, individuals such as Casey persistently reminded the British that intrusions by the Norwegians and Japanese, either through their whaling activities and or claims to Bouvet Island, had to be resisted by flag planting activities or via the dispatch of a major expedition.

The 1930 Imperial Conference codified certain principles for the extension and consolidation of British interests regarding territorial annexation and whaling controls. The British government instructed their Commonwealth partners that the formal act of annexation was no longer sufficient and had now to be supported by measures of formal occupation. The Australian government had been the prime mover in the 1929–31 British, Australian and New Zealand Antarctic Research Expedition (BANZARE) and had subsequently produced new sheet maps for Enderby Land and MacRobertson Land. A memorandum prepared by the British government on the expedition confirmed that the leader Sir Douglas Mawson had been "empowered to take possession of all the areas to which British title had been asserted at the Imperial Conference and of any new territory that might be discovered" (cited in Swan 1961:180 and Bush 1982). The minutes of the first committee also acknowledged the political and economic importance of BANZARE:

> Politically the government was desirous of the flag being hoisted and British title being asserted at all places practicable between King George V Land and Enderby Land. The economic objectives related particularly to the possibilities of whaling being carried out commercially in the region . . . The scientific work would embrace a number of matters comprising . . . the accurate location of the coastline, hydrographic survey work, the charting of islands. (cited in Ferguson 1995:9–10)

In conjunction with the British Discovery research programme, BANZARE was valuable in carrying out marine and oceanographic surveys of the Antarctic coastline. It was also the immediate political context from which an Australian claim to the Antarctic was later presented.

The creation of an Australian Antarctic Territory (AAT) was based on an assertion that the discoveries made by the Australian expeditions between 1911 and 1914 and 1929 and 1930 combined with the whaling activities were of sufficient merit to warrant a claim to nearly a third of the Antarctic continent (see Map 4). The BANZARE had planted the British and Australian flags on MacRobertson Land, Proclamation Island and King George V Land. Moreover, the coastline of the Antarctic from Enderby Land to King George V Land had been sighted and partially mapped (Swan 1961:185–90). The final part of the British procedure, after a public assertion of title and then the formal taking of possession, for acquiring territorial polar claims was the issuing of a letters patent. In February 1933 an Order in Council was issued placing "that part of His Majesty's Dominions in the Antarctic seas which compromise all the islands and territories other than Adélie Land which are situated south of 60 degrees South and lying between 160 degrees East and 45 degrees East". In June 1933, the Governor-General of Australia consented to the act.

The Australian government was anxious that a geographically proximate area was controlled by the British Commonwealth and not by potentially hostile nations such as Norway or Japan. The latter was considered far more threatening by the Australian government not only because of its whaling activities in the Southern Ocean but also due to the long-standing Australian worries of threats from the "North" (Cumpston 1995). The British policy of gradual annexation of the Antarctic continent coincided with Australian ambitions and improved diplomatic communications with Britain through the appointment of Bruce, the former Prime Minister, to High Commissioner in 1933 as well as aiding negotiations over the AAT. Moreover, the role of the polar explorer Douglas Mawson was seminal in raising the profile of the Antarctic for Australian audiences. His involvement in the 1929–31 BANZARE, for instance, ensured that there was sufficient diplomatic momentum for the official declaration. In conjunction with R.G. Casey, these two individuals were critical in restoring Australian post-war polar interests. For the British government, the aftermath of the declaration of the AAT was dominated by two factors: adverse Norwegian reactions and uncertainty over the legal and political significance of American polar expeditions. In January 1934, the Norwegian government sent a memorandum to the Foreign Office contesting the Australian claim and suggesting that a conference be convened between Norway, Britain and the USA to discuss the partition of the Antarctic. Britain rejected this suggestion in favour of a compromise with Norway which had emerged earlier through negotiations with the South African government over not claiming

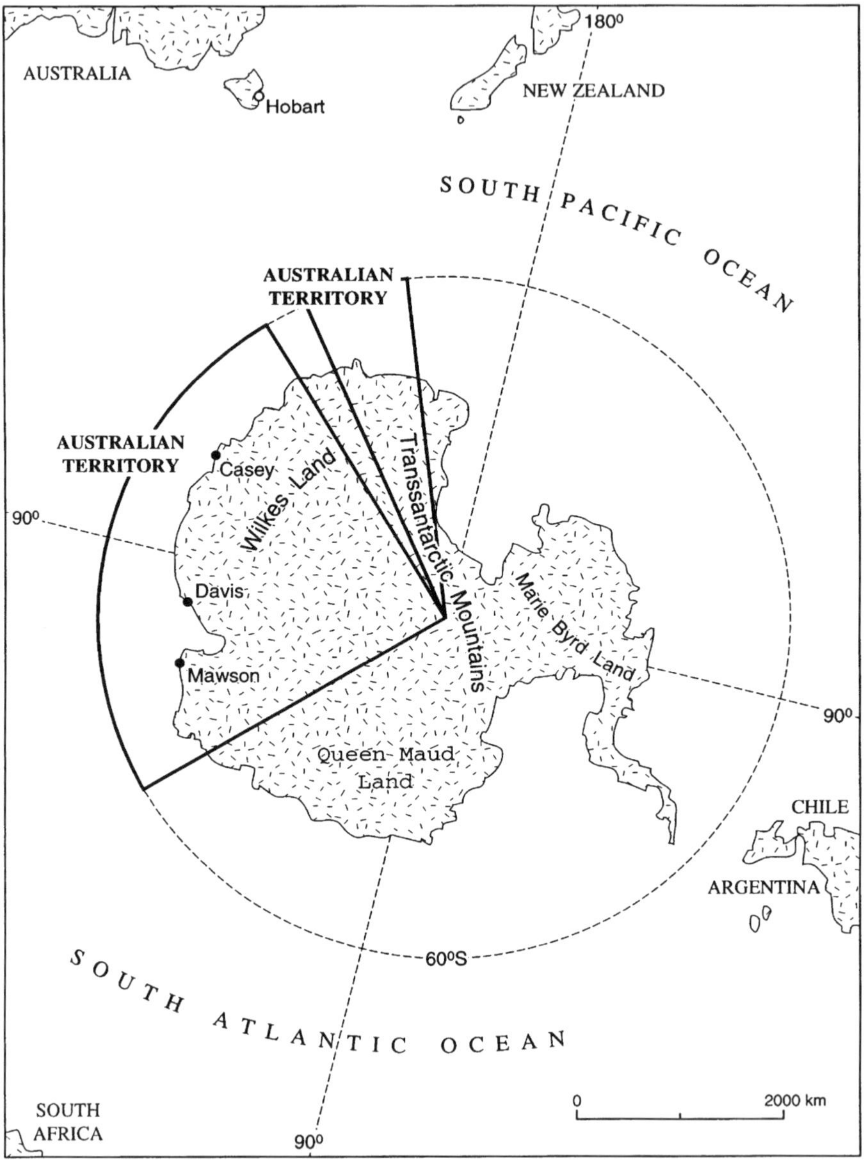

Map 4 Antarctic territory claimed by Australia, and Australia's permanent bases on the continent

a sector around Coats Land. Ultimately, Norway was considered unlikely to contest these claims to the AAT given her willingness to retain good trading relations with Britain and to maintain co-operation with the British Commonwealth over whaling controls in the Southern Ocean.

The 1937 Imperial Conference witnessed intense Australian opposition to the French claim to Adélie Land. Under the chairmanship of Richard Casey, the Committee on Polar Questions concentrated on the French claim and whether it should be resisted. An Australian memorandum on the subject argued that: "The Commonwealth government feels that this narrow French enclave would gravely jeopardise future control and administration of Australian Antarctic Territory as well as complicate further work of exploration." Concern was also expressed that American activity in the Ross Dependency and Norwegian exploration in the AAT made it imperative that the research endeavours of the Discovery Committee were extended and that the Dominions became involved in activities such as meteorological forecasting in the Southern Ocean. From the Australian perspective, uncertainties over the boundaries between AAT and the French claim to Adélie Land were the most pressing. Indeed before the resolution of the problem in 1939, the Australian government had contemplated entering into negotiations with France over the New Hebrides in return for some form of settlement (Hall 1986:33).

The impact of the Second World War on Australian polar interests was profound. The plans for an Antarctic programme were effectively halted by the threat of Japanese forces to the north of the continent. The bombing of Darwin in 1942 deeply influenced the Curtin government. Malcolm Booker cited that the apparent vulnerability of Australia "caused more subtle long term damage [and] created a psychology of dependence on the United States which continued long after the war and still lingers on" (cited in Pilger 1992:155). Ironically, the former Menzies government had already dispatched the Australian armed forces to Europe and the Middle East in defence of Britain's imperial interests. The arrival of General MacArthur in 1942 signalled a new diplomatic and military relationship with the United States. Herbert Evatt, the minister for External Affairs, was most instrumental in not only reorientating Australian relations with Britain (away from blind servility) but also in creating the grounds for a post-war "partnership" with the United States.

Australia's post-war polar activities were shaped by these developments. The Second World War alerted Australian political leaders that their continental island was poorly defended and could be imperilled again by forces to the north. Consequently, defence planning in the post-war years placed great emphasis on securing Australia's northern borders (Ball 1980), whilst southern strategic interests had to be guaranteed so that the armed forces could concentrate their energies on the northern part of the continent. Later fears of tropical and polar encirclement preoccupied sections of the Menzies government as during the mid-1950s, for instance when the Soviet Union announced, in preparation for the IGY, that she would be creating scientific bases in the Australian Antarctic Territory. The activities of German sea raiders in the southern seas had not only illustrated the potential for severe conflict, but it also raised the possibility of submarines, operating under the ice and disrupting the sea lanes of these seas. Australian relations with the United States, therefore, were considered pivotal by the post-war Menzies government, which dominated Australian politics in the 1950s and 1960s. Unqualified support for American actions in Korea in 1950 supported the assertion by Minister of the Interior, Wilfred Hughes, that: "Australia must become the 49th state of America" (cited in Pilger 1992:164). The implications for the Antarctic were also apparent to the Minister for External Affairs, R.G. Casey, who believed that cordial relations with the USA would ensure Australian territorial rights in the AAT in the face of Soviet and Japanese competition.

Australia and the post-war polar revival

The immediate aftermath of the Second World War witnessed a major administrative and logistical revival of polar interests. In a sense such a transformation seemed out of kilter with the grander vision of the then Minister for External Affairs Herbert Evatt. In contrast to the coalition government of Robert Menzies, the Labour government under Ben Chiefly (who had replaced Curtin and Forde Labour governments in 1946) pursued a vision of an Australian nation not dependent on the United States but as a state characterized by non-alignment, liberal democracy and a defender of the Declaration of Human Rights (Cumpston 1995:41). There are some parallels with the political vision of Juan Perón and his attempts to forge a "third approach" to Argentine foreign policy (Melinsky 1978). Evatt, as the first president of the United Nations, had attempted to resist American pressures to incorporate Australia into a Cold War network of defence co-operation and intelligence collection (Ball 1980). The Chiefly government eventually agreed to participate in an "exchange of intelligence" with the United States and the United Kingdom. As was to become abundantly clear later, Australia's Antarctic interests were not immune to sustained American pressure.

In 1947, the Antarctic Division within the Department of External Affairs was created in conjunction with the Antarctic Expedition Planning Committee, which was charged with organizing Australian National Antarctic Research Expeditions (ANARE). This was an administrative innovation and largely due to the recommendations of secret meetings held within the Department of Defence during December 1946. The Department of External Affairs had earlier concluded that: "There appears to have been no active strengthening of our claims since the passage of the 1933 Act, but it should be remembered that the acquisition by the Commonwealth of the ship *Wyatt Earp* and a scientific programme proposed in some detail by Sir Douglas Mawson are evidence of the Australian intention to provide effective occupation." In June 1947, the Australian cabinet agreed that an Australian expedition would be launched in the 1947–8 polar summer season. In August 1947, the government further agreed to the creation of two scientific stations on the Macquarie and Heard Islands. The aim of the ANARE was therefore to maintain Australian interests in the AAT, to deny potential hostile nations access to the AAT, to create a number of permanent bases in the region and to carry out an extensive exploration of the AAT using modern transport and communication technologies (Swan 1961, Hanessian 1965:13). As the Department for External Affairs (DEA) concluded: "The aim of the expedition is to maintain Australian and British interests in Antarctica, interests based on the Australian expedition led by Sir Douglas Mawson and also on the share taken by Australia in the expeditions of Scott and Shackleton" (DEA *Current Notes* 1948:75). The significance of ANARE and the post-war revival of polar interest reflected the changing international politics of the Antarctic, the changing political geographies of the Cold War and the Menzies government's decision to strengthen Australian ties with the United States.

The official declaration of an Argentine and Chilean polar claim in the 1940s placed new pressures on British claims to the FID. Those claims not only overlapped, but they also forced the British to invest in a major civilian programme headed by the Falkland Islands Dependency Survey. A British circular produced by the Commonwealth Relations Office in October 1947 later outlined the implications to Commonwealth colleagues:

> Whilst the UK title to some parts of the Dependencies appears to be well founded, considerable areas are probably still res nullius and, therefore, open to acquisition by any power by occupation . . . We have, therefore, reached [the] conclusion that to maintain our title to the Dependencies we should make protests to Argentina and Chile governments regarding their recent encroachments, whilst [seeking] the judgement of the International Court of Justice. (cited in Dodds 1996c)

The British later advised the antipodean governments that further efforts would have to be made in terms of surveying and permanent occupation if these claims were to be protected from outside powers such as the USA or the USSR. The private expedition of the American explorer Finne Ronne, coupled with subsequent official American programmes such as *High Jump*, had convinced the British government that "unless further steps were taken in the immediate future to strengthen our claims in this area, risk of incidents between rival parties of explorers will be much greater" (cited in Dodds 1996c). Finally, the US State Department issued its first proposals to internationalize the Antarctic in 1948. The Australian government objected to these proposals claiming that their territorial rights would be affected and that there had to be a commitment to exclude the Soviet Union and other interested parties such as Japan. At that stage, the

Department of External Affairs had instructed the Australian ambassador to the United States in December 1948 to reject the idea of a condominium and instead propose the creation of an Antarctic commission which would be responsible for co-ordinating scientific projects and the sharing of weather information. Moreover, the Department for External Affairs wanted foreign activities such as the voyages of Japanese whalers excluded from the Southern Ocean and the AAT because of Japanese wartime activities.

The Cold War dimension was indicative of Australian intransigence over territorial claims and subsequent commitment to ANARE. Australian polar authorities such as Philip Law were acutely aware that the polar continent was not geographically removed from the political dramas of the Cold War (Law 1983). As a senior official in the ANARE section of the DEA, Law had considerable influence over the foreign minister's thinking. In a letter to the secretary of state for External Affairs in Ottawa, the Canadian High Commissioner in London recalled (3 September 1948) a meeting with the Foreign Minister over the US proposals to internationalize the Antarctic:

> Dr Evatt opposes any surrender of Australian sovereignty in her sector because of its potential value for defensive and other purposes. A section of the cabinet led by Mr Dedman, Minister of Defence, prefers control by a condominium because of the new strategic concepts created by the development of long range nuclear weapons. This section of the cabinet is of the opinion that Antarctica should be controlled by friendly, powerful forces just because it is important for defence purposes and a possible site for a rocket borne attack on the southern fringe of Australia. (National Archives of Canada, Department of External Affairs, 1948)

The Department of External Affairs' fears over possible rocket-borne attacks from the Antarctic were effectively mobilized by the Menzies government in order to justify a major investment in a polar and sub-polar programme. In the changing dramas of the Cold War such arguments carried considerable political and popular appeal. Evidence for such a proposition can be found in two places: first, the decision by the Australian government to take over Heard and McDonald Islands from the British in 1948. The transfer of sovereignty between Britain and Australia over the Heard and McDonald Islands heralded a new development in Australian polar participation. The creation of a base on Heard was not only the start of a wider ANARE programme but was also understood as part of a Commonwealth strategy to prevent Soviet or other parties inhabiting or claiming a scattered group of southern oceanic islands. The South African annexation of the Prince Edward Islands in 1948 was also widely understood as part of this grander strategic design. The lack of a suitable polar vessel prevented the Australians from developing plans in the mid-1940s to establish a base on the Antarctic continent, however. Second, the new minister for External Affairs, R.G. Casey, reiterated Australia's condemnation of the American proposals for internationalization:

> The fact is that the US made a proposal . . . in which it suggested a pooling of sovereignty in that territory. That proposal has been considered by a number of British Commonwealth countries, some of which, including Australia, supported the view that instead of pooling sovereignty what was required was a reasoned agreement retaining sovereignty, including Australia's sovereignty over Australia's Antarctic Territory, but with the interested States also agreeing to assist each other in the development of Antarctica. (Parliamentary Debates, 25 Oct 1949)

This position was a crucial element in Australian post-war polar policy. There was never any question of the Menzies government ceding sovereignty either to an international condominium or to the authority of the United Nations. The Australian government had hoped that President Eisenhower would make a claim to the unclaimed sector of the Antarctic rather than press on with a plan for internationalization. Douglas Mawson in correspondence with the American polar geologist Lawrence Gould is alleged to have noted that:

> I am still sorry that America has not laid territorial claim to the big unclaimed central Pacific sector. There would be no difficulty with New Zealand, I am sure, having it extended into the Ross Ice Shelf. No doubt also territorial adjustments could be made elsewhere. What seems to me important in such a move is that the Antarctica ice continent if tied up to a limited number of sovereignties of good repute and mutual regard could then be administered on good lines conjointly for the benefit of all and especially of the inhabitants (penguins, birds, seals, etc.). If every nation had a hand in it, some of the life there may soon become extinct. (Gould 1958:32)

America's deliberations as to whether to press a territorial claim to the Antarctic continent were subject to numerous internal reviews by the Department of State in the late 1940s and 1950s (Klotz 1990). In one sense, Menzies's inability to persuade the Americans to declare territorial rights over the unclaimed sector in the Antarctic was indicative of Australia's dependent position *vis à vis* the United States. Later events relating to the Soviet decision to create a number of bases in the AAT during the IGY revealed Australia's sense of vulnerability and fears of a communist annexation.

Australia's reaction to India's plan to internationalize the Antarctic under the control of the United Nations in 1956 exposed Australia's hostility to internationalization and the diminution of its territorial rights. The Minister of the Navy, O'Sullivan noted that, "The Australian government can see no reason for placing the Antarctic continent, including the AAT in particular, under the control of the United Nations" (Parliamentary Debates, 22 May 1956). The Australians were adamant that their sovereign rights in the AAT were non-negotiable. If the Americans could have been persuaded to have pressed a formal claim, it was hoped that their plans for an international condominium would alter from suspending sovereignties to acknowledging that only claimant states could decide the fate of the Antarctic continent. This last point too had implications for the possible involvement of the Soviets in the polar continent. Richard Casey, as Minister of External Affairs, repeatedly expressed concerns that: "The Australian Antarctic sector is of vital importance to Australia. For strategic reasons, it is important that this area, lying as it does close to Australia's backdoor, shall remain under Australian control" (Australian Archives, Department of External Affairs 1957).

Australian representations of the Antarctic and the Southern Ocean were thoroughly imbued with the drama of the Cold War and longstanding concern over territorial sovereignties. Peter Beck has argued that "In general the southern continent was interpreted through strategic/Cold War spectacles, as evidenced by the desire to settle the Antarctic question without reference to Moscow . . . Similar perceptions [as compared to the USA] influenced official thinking in other countries, particularly those located in the southern hemisphere" (Beck 1990c: 200 and 203). The role of cartographic imagery was also notable during the 1950s as maps were employed to highlight the geographic proximity of the Antarctic continent to the Australian mainland. The description of 3,000 miles of ocean and ice as a "back door" demonstrates how the perceived significance of distance was changing in the advent of the nuclear age and the intercontinental bomber.

The key moment in the history of ANARE, however, was the 1954–5 season when the Australian party established a base on MacRobertson Coast, Antarctica. It was named Mawson base in honour of the famous British-born explorer Sir Douglas Mawson and, as the first permanent base, it became the centre for Australian Antarctic activities. Simultaneously, the Menzies government introduced legislation confirming that the AAT was subject to the laws of the Australian Capital Territory (Joyner 1990:54). As Walter Sullivan commented: "It was as though the District of Columbia had been enlarged to include Alaska, and was further evidence of Australia's determination to maintain her foothold in Antarctica" (Sullivan 1957:123). It was recognized that effective occupation combined with administrative procedure was considered crucial to strengthening polar territorial claims (Heron 1954:661).

This mounting interest in consolidating Australian polar claims coincided with British attempts to launch a Commonwealth Trans-Antarctic Expedition (TAE) in response to fears not only of Argentine encroachment but also New Zealand's apparent ambivalence over its administrative responsibilities in the Ross Dependency. This was deeply ironic given that British activity had been greatest in the Ross sector. In 1954, the British raised with Commonwealth colleagues the possibility of increasing polar activities across Commonwealth territories. The Note on Future UK Policy in the Antarctic argued that: "We are impressed by the scientific and political value and the prestige of such a new Antarctic journey . . . It would, we are advised, have little value towards strengthening our claims to sovereignty . . . but it would demonstrate to the world our leadership in the area (British Archives, Foreign Office 1954).

The Australian response to the proposed Trans-Antarctic Expedition was undoubtedly muted compared to the financial and political response of South Africa and New Zealand. The Menzies government argued that Australian commitments in the AAT were already substantial and precluded financial and logistical investment in the TAE. The Australians later pledged £20,000 which was then considered a generous contribution given that over £200,000 was being spent on developing bases in the western sector of the AAT. Within Australia, domestic support for political claims to the AAT was widespread. Newspapers such as the *Melbourne Herald* were ardent supporters of the Australian presence and the need to prevent the "Reds have[ing] a toe hold in Antarctica". Hanessian reported that:

> Australia's active Antarctic policy is based on considerations of security, prestige, science and possible economic gain. It has strong bipartisan support in Parliament and in the press. Popular interest in Antarctica is widespread – the publicising of recent Antarctic activity has created an impression of the existence of a tangible and potentially valuable Australian interest in the region, an attitude that has not been denied by the Government. (1965:14)

A notable development of the mid-1950s was the manner in which Australian ministers were consulting with North American colleagues over the consolidation of territorial claims in the polar regions. In a meeting on 9 September 1955 with Lester Pearson of the Department of External Affairs, Richard Casey outlined his fears of Soviet and American involvement in the AAT. Pearson recorded that his Australian counterpart was increasingly worried that the United States would mount a legal and political challenge to the AAT. Casey was anxious to know how the Canadian experience in the Arctic especially over co-operation in the face of fears of Soviet activity in the Arctic region, had fared. Pearson later suggested possible UN involvement as a potential solution to rising tension in the Antarctic region, but Casey rejected such a proposal and was also to voice bitter opposition to the Indian proposal to internationalize the Antarctic and American plans to institute a condominium. Indeed there were Australian concerns that American intransigence over the latter might be encouraging the Soviet Union to consider pressing a formal claim (Toma 1956:624).

Australian unease about Soviet encroachment remained in the run-up to the International Geophysical Year (IGY). In a meeting of the SEATO parties in March 1957, Richard Casey warned the American Secretary of

State J.F. Dulles that the activities of the Soviet polar scientists would have to be monitored closely. The Australian government once again expressed the hope that the Americans might enter into an informal alliance (given the existence of SEATO and ANZUS) in order to prevent the Soviet Union from establishing a permanent and threatening presence on the Antarctic ice (T. Miller 1977). The Soviet Union had established bases at Komsomolskaya and Vostok in the AAT during the research period, whilst the Australians had created a new base in the Vestfold Hills called Davis in 1957. Until the ending of the IGY, the Australian government had been hostile to any international agreement altering the territorial status quo. In a meeting on 27 August 1957 with the Canadian High Commissioner to Australia, Richard Casey had concluded that:

> The Australian interest in the Antarctic is primarily political and strategic . . . it was essential that no publicity be given to these discussions until the future intentions of the USSR were clearer. As for a condominium, Mr Casey's view was that no such arrangement would be worth while unless it provided for demilitarisation of the zone. In that case it would be possible to include the USSR in the international body, otherwise the strategic concerns of Australia would be considerable. We do not want the Russians to mount installations in the Antarctic from which they could drop missiles on Melbourne or Sydney. (National Archives of Canada, Department of External Affairs 1957)

In the final months preceding the Washington Conference in October 1959, the US government arranged a number of meetings with Commonwealth nations. The Australian position with regards to possible changes in the international politics of the Antarctic was dominated by a fear of the ever-present Soviet IGY bases in Antarctica as well as proposed plans of some governments to involve the UN in the management of the continent. In March 1958, Menzies discussed with Prime Minister Nash of New Zealand the issue of internationalization and warned that Australia would only accept an advisory commission and not full-scale UN involvement.

Australian objections to New Zealand's plans for internationalization emerged again in September 1959. Menzies dispatched a letter to Nash dated 7 September 1959 stating his objectives for the forthcoming Washington Conference. Four conditions were cited: the demilitarization of the Antarctic, protection of Australian territorial claims, avoidance of an international authority and a refrain from devising measures relating to the exploitation of natural resources. There was widespread concern that after the IGY the Soviet Union had retained its polar bases located as they were in the AAT, for ulterior political motives. As the Adelaide *Advertiser* noted on 12 February 1957, "International communism would then be in a position to prepare a pincer movement against this continent . . . What will the Australian government do if the Russians do not leave Antarctica at the end of the year?" As if to reiterate the importance Australia attached to the AAT, Menzies also announced that the polar programme for the

1959–60 season would involve personnel being based in Davis, Wilkes (inherited from the Americans), Mawson and Macquarie. Planes and snow tractors were to be used in a major mapping and geological survey of MacRobertson Land and Enderby Land. Moreover, since 1954, legislation had ensured that the laws of the Australian Capital Territory were now applicable to the AAT. The legal, political and geographical "occupation" of the AAT, although incomplete, was considered to be of paramount importance on the eve of the Washington Conference (Swan 1961, Price 1962).

Australia and the 1959 Antarctic Treaty

In his opening statement to the Washington Conference, the head of the Australian delegation Richard Casey reminded his audience of Australia's special geographical relationship to the polar continent:

> The Antarctic's geographical relationship with the Australian continent means that it is of special importance to us. Australia's southern coastline of 2,000 miles runs parallel to the Antarctic coastline *only* 2,000 or 3,000 miles to the south. The weather of Australia is greatly affected by Antarctic meteorological conditions . . . Australia has a long history of exploration and scientific activity in the Antarctic and the Australian Antarctic Territory makes up almost one third of the continent . . . In view of our efforts in the Antarctic over several decades, I feel that I am justified in claiming that Australia has made an important contribution to the opening up of this difficult and inhospitable continent. (Australian Archives, Department of External Affairs, 1959)

The Australian delegation was regarded by other Commonwealth representatives as the most militant over their claims to an Australian Antarctic Territory. John Heap, the former head of the Polar Regions section in the Foreign Office, recalled the "fear of chaos" as the 12 parties negotiated the legal and political foundations of the Antarctic Treaty (cited in Beck 1986). However, Australian concern over their sovereign rights in the Antarctic was replicated by other southern hemispheric states such as Argentina and Chile. Australia's geo-strategic position was taken to be indicative of their concern over sovereignty and unrestricted access to the polar continent.

The formulation of Article IV was undoubtedly the central achievement of the Antarctic Treaty. Recognizing that the claimant states were not prepared to cede territory and that the non-claimant states would not recognize these claims, the delegates formulated a text which effectively suspended the issue of territorial sovereignties. Australian objections were overcome because their claims to the AAT did not have to be formally renounced. Ironically, Commonwealth observers later conceded that the most intransigent nation over polar claims had been France rather than southern hemispheric nations such as Argentina or Australia. The Canadian Embassy in Washington noted that the French delegation had arrived at the Conference warning that their sovereign claims to Adélie Land would not be compromised by any new legislation. Initial Australian hopes of settlement over territorial claims, it was reported by the Canadian observers, were shattered by the French "bomb-shell". The Soviet delegation later suggested that: "The French be allowed to talk themselves out until they see eventually that they are one against eleven." This apparent French intransigence was overcome after much political lobbying by the American, Soviet and British delegations.

Another complicating issue for the Australian delegation during the Washington Conference was the denuclearization of the Antarctic continent. Richard Casey, the Australian Foreign Minister, argued that "the southern hemisphere [was] a 'special case' on account of its relatively low radioactivity levels" (Beck 1990c:206). Perversely, however, the Menzies government in 1959 had just provided shipping and nuclear testing facilities to the British on Christmas Island in the North Pacific (Lee 1995). Notwithstanding these inconsistencies, the Australian delegation was joined by the South American and South African states in demanding a complete ban on all nuclear testing. Moreover, Richard Casey argued that the Australian government would not sign an Antarctic Treaty which excluded this provision in spite of opposition from the Western nuclear powers (the USA and the UK). In November 1959, the Australian delegation provided a compromise clause which suggested that only nuclear activity with the intent of promoting scientific investigation would be tolerated, whilst the Soviet Union had expressed demands for a total ban on nuclear activity in the Antarctic region (see Beck 1990c:208). It was not until late November 1959 that Article V materialized in the form of a complete ban on any nuclear explosions in the Antarctic.

In the aftermath of the Washington Conference, the Australian delegation recorded satisfaction with proceedings, especially over Article IV concerned with territorial claims and Article V relating to the prohibition of nuclear explosions and testing on the polar continent. The official report of the Australian delegation concluded that Australia had secured a number of benefits:

> We have obtained a treaty which safeguards our backdoor . . . It ensures that the Antarctic will not become an arena of great power competition . . . It protects our national position in the area by maintaining the status quo in regard to claims . . . During the conference Australia was able to make further progress towards establishing itself as the leading Antarctic power in the southern hemisphere. (SPRI Archives 1961)

In his opening address to the First consultative meeting of the ATS held in 1961, the Australian Minister for External Affairs Robert Menzies reiterated the above:

> Consequently, all the nations concerned have agreed that this is not a place of war, but that this is a place of scientific research, of study, of enlarging the boundaries of knowledge, of friendly co-operation . . . The third thing is that under the Treaty we have agreed to set aside the argument about territorial claims. Nobody abandons his own. We have made territorial claims in the Antarctic – quite expansive ones . . . We have said to each other, "look there are more important things than arguing about our rights". (ATCP 1961)

From the Australian perspective, it was considered imperative that the Antarctic should be declared a zone of peace and co-operation. With the effective demilitarization of the polar continent, the Australian government could concentrate on more pressing strategic concerns to the north of the country which had preoccupied Australian political leaders throughout the 1950s. Fears had also been expressed that either Soviet-inspired communism might overcome the governments of South East Asia and/or that Japan might begin to exert herself in the South Pacific to the detriment of Australia. The "question of Antarctica" was, for Foreign Minister Casey, one piece in a complex strategic puzzle shaped by the geographical and political currents of the Cold War.

Protecting strategic "back doors": Australia as a regional power

Australian concerns over the Antarctic and the Southern Ocean need to be understood within a broader regional setting (see, for example, Lee 1995). In the immediate post-war period, Australia had become immersed not only in defence organizations such as ANZUS but also assumed the role of a regional power in the South Pacific. In 1950, the Liberal government offered to take over British administration of the New Hebrides and later engaged in discussions over the Solomon Islands. In the case of the former, an earlier offer in 1937 had been rejected by the British government because of Australia's substantial territorial responsibilities in the Australian Antarctic Territory and in the South Pacific. Interest in the South Pacific islands of the New Hebrides was largely shaped by the recent experiences of the Second World War and the usage American troops had made of their harbours. Australian strategic concerns in the South Pacific had therefore become thoroughly immersed in the drama of the Cold War. The recent communist take-over in China and the rise of insurgencies in Indo-China had persuaded the Liberal government that a Pacific pact with the United States, or possibly Britain, in order to guard against either communist expansionism or a resurgent Japan would be expedient. As David Lee has noted: "The Menzies government endeavoured to throw off Australia's dependence on the sterling area and to forge with the United States a new economic relationship which would allow both rapid economic development and defence preparations" (Lee 1995:108).

The decision to conclude a defence pact with the United States and New Zealand in 1951 caused considerable concern in Britain. The ANZUS pact excluded Britain from a major defence agreement between two Commonwealth partners yet left Britain responsible for the defence of the Borneo territories and Pacific islands such as Fiji. The Menzies government issued assurances to the British government that Australia was not seeking to remove Britain from Australian security considerations somewhat prematurely given that Britain had permission to carry out nuclear testing in the interior of Australia. Insecurities in the South Pacific were matched to an extent in the Antarctic and the Southern Ocean. Successive Australian governments were convinced that Australia lacked sufficient resources and manpower to defend its strategic interests. Richard Casey noted in 1952 that: "Need for additional administrative services in our existing Australian territories has been growing, and we are feeling that our resources in experienced administrative personnel cannot keep pace with our requirements". The Menzies government later withdrew its offer to administer the New Hebrides due to fears of the aforementioned administrative over-commitment and the rising costs of administering New Guinea and the Australian Antarctic Territory.

Anglo-Australian relations over defence and security matters during the 1950s were undoubtedly tense. In the South Pacific, Australian ministers were concerned over Japanese investment in the New Hebrides and British indecision over her Pacific possessions. Bitter memories of Japan and the Second World War made the Menzies government sensitive to any political or strategic change. The ANZUS pact had boosted Australian strategic confidence in the South Pacific, and the SEATO treaty in 1954 had reduced fears of a communist take-over of South East Asia. In terms of new administrative responsibilities, Australia later took over the Cocos and Keeling Islands from the British in 1955. In the Antarctic, however, Australian fears were dominated by the presence of the Soviet Union. The latter had created a number of polar bases in the AAT, and concerns were expressed in Canberra that the Soviets might launch a missile attack on Melbourne or Sydney. Danger for the Menzies government was to be found, therefore, both north and south of the Australian continent.

The significance of these observations was apparent given that successive Australian administrations in the post-war period have had a distinct geographical sense of Australia as a nation with specific regional interests in the South Pacific, South East Asia and the Antarctic. The 1959 Antarctic Treaty effectively ensured that the polar continent would be a demilitarized space and that existing claims to territory would not be subject to "decolonization". As the Australian delegate D. Hay reminded colleagues at the second ATCP meeting in 1962:

> My government, has from the start, attached significance to the objective of the Antarctic Treaty. After all, Antarctica is geographically near to Australia. It is not surprising that the provisions of the Treaty for international scientific co-operation, as well as for demilitarisation and for the setting aside questions of territorial claims are highly valued in Australia. (ATCP 1962)

Nonetheless, successive governments have consistently expressed the view that Australian involvement in the Antarctic and the southern seas is linked to wider strategic, meteorological and defence concerns. As with the other southern hemispheric rim nations, Australian claims to the Antarctic were considered an important expression of those interests.

Australia and resource interests in the Antarctic and Southern Ocean

Following the 1959 Antarctic Treaty, Australia devoted considerable resources to maintaining a scientific programme in the Antarctic and Southern Ocean and emerged as a major player in co-ordinating the business of the Antarctic Treaty Consultative Parties. The first Consultative meeting in Canberra in 1961 was a milestone for the ATS (Australia was disappointed not to host the Antarctic Conference) as delegates negotiated procedures, regulations and agendas. Rules of procedure and 25 recommendations were adopted during the meeting (Chaturvedi 1990:98). More significantly, the Canberra meeting also tackled the issue of conservation of living resources, especially those animals and plants indigenous to the Antarctic. The impetus for these discussions was the UK government which had argued that the development of agreed measures should be an important feature of future international co-operation. The eventual adoption of the Agreed Measures for the Conservation on Antarctic Fauna and Flora (1964) was achieved because it did not compromise formal sovereignties and was geographically restricted to the land and the ice shelves. The high seas were not formally included in the remit of the Agreed Measures as with the Antarctic Treaty. The subsequent legislation implementing the Agreed Measures included the high seas in contrast to British and New Zealand legislative packages (Bush 1982:147). In the 1950s and 1960s, Australian governments adopted the view that the high seas should be included in the remit of the Antarctic Treaty, not least because of their concerns that the entire region below 60 degrees south should remain a demilitarized space.

The most pressing concerns for the Australians in the 1960s and 1970s revolved around the issues of territorial sovereignty, the avoidance of armed conflict and resource exploitation. The legal and political protection of the AAT had remained of paramount importance in spite of the caveats of Article IV of the Antarctic Treaty. On the one hand, the Australian government had insisted that their sovereignty over the AAT could not be compromised in spite of the introduction of agreed measures and conventions for the protection of living resources. Yet on the other hand, the Australians, as with other southern hemispheric nations, had invested heavily in research on communication technologies. The potential for military application rather than strictly scientific benefit has already been noted by some polar authors (eg Hall 1986). During the 1960s, Australia was active in establishing telecommunications on the Antarctic continent and was preoccupied with fears that other weather or radio stations might be formally recognized by the World Meteorological Organization as official stations within the AAT.

It has been persuasively argued that the Australians retain a considerable concern for their resource rights over the AAT and surrounding oceans. During the negotiations over the CCAMLR in the 1970s, the Australian government argued that their rights over the seas around the AAT and islands such as Heard should not be compromised by the provisions of the Convention for marine conservation. The declaration of an Australian Fishing Zone (AFZ) in September 1979, however, established a new legal requirement for foreign fishing boats to obtain licences before fishing in AFZ waters. This type of objection was also voiced by the French delegation with reference to their possessions in the Kerguelen and Crozet islands. Consequently, Article IV in the CCAMLR reaffirmed the spirit of Article IV in the 1959 Antarctic Treaty which protected the existing legal sovereignties of the claimant states (Joyner 1990b:315–16). The final convention determined that the area covered by the CCAMLR would constitute a definitive ecological regime without compromising the coastal state jurisdiction in the region (Joyner 1990b:316). This was considered to be crucial to allay fears that the Soviet Union was intent on expanding its harvesting of krill, fish and whales. The oceans and seas around the AAT were subjected to substantial Soviet and Japanese fishing whilst a new base, Leningradskaya, was constructed in the eastern portion of the AAT in 1970. Moreover, the Gorton government expressed reservations about the growing Soviet influence in the Indian Ocean *per se* (Cumpston 1995).

The question of mineral exploitation in the 1970s was a most pressing issue for Australian and other southern hemispheric governments because Australia was: "[an] energy deficient nation who saw in the Antarctic the last hope of satiating their thirst for oil. An Australian spokesman has been quoted as saying, 'the treaty did not freeze territorial claims; it merely put them on the back burner to keep warm' " (Chaturvedi 1990:138). Since the 1970 ATCP meeting in Wellington, the Australian delegation had been adamant that any mineral resources found within the AAT belonged to Australia and that a moratorium on mining would be temporary and would also depend upon others respecting Australian sovereignty. Up to the 1981 Buenos Aires ATCP meeting, successive Australian delegations had reiterated their belief that the mineral resources under the AAT could not be developed by technologically more advanced nations such as the United States. A major stumbling block throughout the 1970s was the need to devise a mechanism whereby the sovereign rights of claimant states were recognized whilst at the same time the non-claimant states were not placed in a position of having to accept that substantial economic benefits would be derived in favour of the claimant states. As the leader of the Australian delegation to the 1981 ATCP meeting Keith Brennan noted:

> An important issue which has arisen recently is the increasing interest in Antarctica from a number of quarters including non-Treaty nations, and private environmental and commercial organisations. This development has been largely stimulated by greater interest in Antarctica's resource potential and by current world economic conditions which tend to encourage the search for new resources. (ATCP 1981)

This basic stance was to be repeated during the CRAMRA negotiations in the 1980s when Australia's position on resource management was governed by concerns that Australian sovereignty might have been compromised by unregulated exploitation. Moreover, concerns over the consensus tradition of the ATS also led to fears that the basic division between claimant and non-claimant states within the ATS would provoke a fracturing of this international system. Paradoxically, therefore, Australian interests were concerned with maintaining territorial rights in the AAT whilst at the same time trying to ensure that the ATS did not split over resource questions.

Significantly, Australian geopolitical reasoning changed in the late 1970s and early 1980s. Before the mineral negotiations, Australian public statements either to the ATCP meetings or in Parliament were characterized by a naturalized geopolitical reasoning which stressed that Australian sovereign rights in the AAT were derived from, amongst other things, their geographical proximity to the Antarctic. The enhancement of sovereignty was the number one priority of the Australian polar programme. In the 1980s, however, there was a shift away from the regular pronouncements on sovereignty to the promotion of Australia's role within the ATS and the need for vigilance in the face of new challenges to the legitimacy of that system. The negotiation of CCAMLR stimulated this transformation as did the appointment of Richard Woolcott to the chairmanship of the New York Group of the Antarctic Treaty Parties between 1982 and 1988. The Australian *Foreign Affairs Record* (AFAR) concluded in 1981 with respect to the former:

> With the conclusion of a marine living resource convention, the Consultative Parties have once again focused their attention on the much more difficult problem of drawing up a regime (plan) to govern the possible exploration and exploitation of Antarctic mineral resources . . . They [ATCPs] are particularly concerned that unless regulatory frame-works can be established within the

> Antarctic Treaty System, unregulated activity will take place which could damage the fragile Antarctic environment and at the same time threaten the viability of the Antarctic Treaty System. (AFAR 1981:5)

The negotiations over CRAMRA began in earnest during 1982–3 and culminated in the 15th ATCP meeting in Paris during October 1989. Whilst it has become increasingly common for commentators to draw attention to the sudden Australian rejection of CRAMRA in 1989, the process of disenchantment had in fact been established relatively early on in these negotiations (Bergin 1991, Elliot 1994a, Morgan 1996). The environmental protection of the Antarctic became an increasingly important theme for Australian delegations during the 12th ATCP meeting in Canberra (1983), the 13th ATCP meeting in Brussels (1985) and the 14th ATCP meeting in Rio de Janeiro (1987). The Australian delegations were further aware that a number of other events such as the Malaysian challenge in the United Nations, the accession of new members to the ATS such as China and India coupled with objections because of the Pinochet military regime to Chile's hosting the next ATCP meeting in 1989 were placing the ATS under considerable strain (Woolcott 1984, 1990). As Bill Hayden of the Department of Foreign Affairs acknowledged at the 1983 meeting in Canberra:

> The Antarctic Treaty has provided a framework for preserving and protecting the Antarctic environment, a matter of high priority for my government . . . One of the principal reasons why Consultative Parties have commenced negotiations on a minerals regime is in order to ensure that no unregulated minerals activities takes [*sic*] place, including even prospecting, which could be harmful to the Antarctic environment . . . In our view it will be particularly important in the months and years ahead for the Treaty parties not only to continue to act with the cohesion which has characterised their co-operation under the Treaty, but also to show flexibility to achieve the common goal of preserving the Antarctic Treaty. (ATCP 1983:51–2)

Whilst the Department of Foreign Affairs supported the principles underlying the CRAMRA proposals, it was increasingly apparent that the minerals negotiations had provoked concern within the Hawke government over the future stability of the ATS. During the 1985 and 1987 ATCP meetings, the Australian delegation raised the possibility of declaring so-called "Managed Areas" in the Antarctic. These would involve developing new administrative structures for the management of special areas of the continent. In conjunction with proposals for a code of conduct for waste disposal and bureaucratic reform of the ATS, the Australian government appeared to be moving away from a narrow definition of national interest based on sovereignty towards a broader agenda encompassing the future workings of the ATS and environmental protection (Morgan 1996).

This shift of interest in the management of the Antarctic should not be exaggerated as the transfer of the Antarctic Division from the Department of Science to the Department of Arts, Sport, Environment, Tourism and

Territories in July 1987 indicated that the division became more closely aligned with the views of the Department of Foreign Affairs (Bergin 1991:232). The latter was supportive of the CRAMRA proposals for the regulation of mining coupled with the environmental protection of the Antarctic. As the *Annual Report* of the Department of Foreign Affairs and Trade (DFAT) concluded in 1987–8, "The Convention [CRAMRA] was designed to set in place an effective international system of control well in advance of any minerals activity. It includes strong provisions to safeguard the Antarctic environment" (DFAT 1987–8:80). Although by the time the Convention was opened for signing in June 1988, the ATCP had endured considerable controversy surrounding the minerals debate. The activities of Greenpeace in exposing French runway plans in the Antarctic coupled with a highly effective media campaign had begun to shift Australian public opinion firmly against CRAMRA. The Australian delegation at the United Nations under Richard Woolcott had chaired the ATCP Co-ordination Group and had been charged with the responsibility for responding to Malaysian and other criticisms of the ATS. The failure to agree conservation measures under the CCAMLR during the 16th ATCP meeting in Rio de Janeiro did not assist the negotiating atmosphere.

Australia, CRAMRA and the 1991 Protocol on Environmental Protection

The Australian decision to reverse her position on the CRAMRA was dramatic considering the earlier stated views of interested government bodies such as the Department of Arts, Sport, the Environment, Tourism and Territories (DASETT). In their *Report* on the Australian Antarctic Science Programme, the DASETT had concluded in 1988 that Australian interests in the Antarctic would need to increase in scope: "In order to sustain its claims to a substantial role in the management of mineral activities in the AAT, Australia should have a scientific program which will provide the information and understanding required for effective evaluation, management, and regulation of mineral resource exploration and exploitation in the region" (DASETT 1988). The Australian Antarctic Scientific Advisory Committee concluded that the polar programmes should concentrate on a number of objectives which included "the optimisation of the special opportunities . . . from the Antarctic continent and Southern Ocean and their proximity to Australia" as it related to resource evaluation (DASETT 1988:35).

Australian opposition to the CRAMRA proceedings in 1989 highlighted the profound uncertainties that existed over mining and mineral exploitation and whether these could be accommodated within the ATS or any other legal regime. The decision to withhold signature from the minerals convention effectively destroyed the fragile consensus of the ATS. CRAMRA required unanimous agreement amongst the claimant states for that convention to be enforced (Joyner 1990:56). Richard Woolcott, an Australian foreign affairs official, in a speech to the Australian Institute of International Affairs in August 1990 argued that:

> We have taken the lead, initially against the opposition of many important countries, to prevent once and for all any future mining and oil drilling in Antarctica. It is planned to turn this magnificent and fragile wilderness continent into an international nature reserve and land of science. Australia's Antarctica initiative seeks to build on the unique achievements of the Antarctic Treaty. We recognise that on this issue we have to set our sights high, and that the road ahead will be neither easy or quick. (Woolcott 1990:48)

It has been suggested that the sudden rejection of the CRAMRA proceedings by the Hawke government was related to the growing strength of the green vote and the specific election of green representatives as a result of the Tasmanian state elections in May 1989 (Bergin 1991, Elliot 1994b, Chaturvedi 1996). Given that these new members were considered essential to Hawke's minority government, demands for evidence of Hawke's environmental commitment heightened. The rejection of CRAMRA was motivated, therefore, primarily by domestic political considerations notwithstanding the growing importance of environmental affairs in international politics more generally (Bergin 1991). Other authors have pointed to the lobbying power of the Australian mining corporations (such as the Australian Mining Industry Council) and their objections to the possibility of subsidized polar mining by non-Australian firms at the expense of Australian companies (R. Dunn 1989). The Council, however, had been supportive of CRAMRA. Indeed, the Australian delegation to CRAMRA later argued, unsuccessfully, for an anti-subsidy clause in order to protect their domestic mining interests. Finally, some observers have pointed to the importance of recent environmental disasters such as the grounding of the Argentine polar supply ship *Bahia Paraiso* in 1989 and the large oil spill caused by the *Exxon Valdez* in the waters of the Arctic. Media images of the apparent fragility of the polar regions to environmental disasters heightened public and political anxieties over the possibility of mining in the Antarctic. Ecological movements such as the Australian Conservation Foundation, under the leadership of Philip Toyne, succeeded in mobilizing substantial public opposition to CRAMRA in Europe, North America and Australia (Darby 1994:109).

Australia's objections to CRAMRA (soon to be joined by France) were rejected by the British government which had argued that:

> The Australian case rests on an insubstantial assertion that Antarctic mineral activity is now suddenly environmentally unacceptable. The idea of maintaining the Antarctic as an unsullied paradise free of environmentally damaging mining activities is attractive . . . But the purpose of the convention is precisely to avoid

environmental damage from mining in Antarctica. Even if an outright ban on mineral activity could be negotiated – we do not believe that it could – it would be extremely unlikely to work. In a world of increasing population and increasing pressures on a finite resource base prospecting for minerals would continue. Instead of such prospecting being carried out openly within Antarctica and subject to proper control, as laid down in the convention, it would be done covertly under the guise of scientific research. (cited in Chaturvedi 1994:20)

In August 1989, the French and Australian governments issued a joint statement detailing their objections to the Convention on Antarctic minerals. They argued that the environmental fragility of the Antarctic would not tolerate mining or mineral exploration. Instead of CRAMRA, the two states proposed a convention on the environmental protection of the polar continent. Sir Ninian Stephen, then Australian Ambassador for the Environment, was dispatched to Moscow, Brussels, Bonn, London, Paris and Wellington in order to advance the Franco-Australian initiative (DFAT 1989–91 *Annual Report*:107). Far from being a completely disinterested diplomatic initiative, Australian claims to the AAT might have been "compromised" by a ratified CRAMRA if other states had explored the mineral potential of the AAT (Joyner 1990:56). The proposal for an environmental convention was put before the 15th Consultative Party Meeting held in Paris in October 1989, and the ATS membership later agreed to convene a special meeting in Chile for the following year to consider the Australian–French proposal.

Ironically, Woolcott's appeals for a new global partnership and the prohibition of mining preceded a 1989 Parliamentary Report on Antarctic tourism which reinforced the notion that the Antarctic continent was under the jurisdiction of property-owning nation-states. Australian sovereignty was once again reasserted over the Australian Antarctic Territory.

Australia is one of seven states which claim territorial sovereignty over parts of the Antarctic continent . . . In Australian law the position is quite clear – the AAT is an Australian territory in every way and is of the same status, for instance, as Norfolk Island or Christmas Island. Therefore, Australian law applies subject to relatively few qualifications . . . The regulation of tourism under Australian law and governmental policy would be an exercise of our sovereignty over the AAT. (Australian House of Representatives 1989, cited in Beck 1994c:3)

The Australian Antarctic programme remained wedded to the firm belief that the administration of the AAT remained a fundamental policy objective of the Antarctic Division of the Department of Education, Sport and Territories (DEST) (Morgan 1996). In support of this particular objective and others such as environmental management, the Department of Foreign Affairs created an Antarctic Foundation in November 1990 to support informed debate on the Antarctic through the establishment of conferences, seminars and information networks (DFAT 1989–91 *Annual Report*:106–7).

The actual discussions surrounding the formal rejection of CRAMRA and the subsequent proposals for a Protocol of Environmental Protection are reasonably well known (Beeby 1991, Bergin 1991, Chaturvedi 1996). Australia's proposals for a new initiative on environmental protection was considered in an ATCP meeting in Paris in October 1989 and then later in Vina del Mar in December 1990. In his opening address to the Paris meeting, the Australian representative, Alan Brown, reiterated his country's position that the joint proposal with France provided: "the outline of elements which would establish broad principles, identify action required and put in place machinery which will be essential if a comprehensive regime for the protection of the Antarctic environment is to be effective" (ATCP 1989:122).

The negotiations over the Australian–French proposal were undoubtedly problematic given the opposition of southern hemispheric nations such as Argentina as well as the USA, the UK and Norway. The Australian foreign minister, Gareth Evans, admitted in Parliament in December 1990 that the Vina del Mar special meeting had been difficult:

> The Treaty parties adopted as a basis for negotiation a text which states that, and I quote: Any activities relating to mineral resources, other than scientific research, shall be prohibited . . . That text was slightly unusual to the extent that it was followed by three dots, which rather conveys the notion that at least some nations do not necessarily accept that being the last word on the subject. (DFAT 1990 *Monthly Record* December:884)

Moreover, important networks such as the Correspondents Group within the ATS also declined in significance given the differences of opinion between the USA, New Zealand, the UK and Australia. Australian officials later admitted that Australia had been isolated during the Vina del Mar.

The following two years witnessed growing support for an environmental convention after a series of important meetings in Madrid and Bonn during 1991. The signing of the Protocol in Madrid during June 1991 confirmed that a crisis over CRAMRA had been averted. The Australian delegation to the 16th ATCP meeting in Bonn concluded that consensus within the ATS had returned on the mining and mineral exploration issue. The draft protocol reflected a delicate balance of interests which ensured that whilst mining was banned for at least 50 years, a new legal regime for the regulation of mineral exploitation would have to be in place before then. Although there have been many critical commentaries of the final version of the Protocol, the diplomatic and political significance of this Protocol must not be underestimated (ASOC 1991, Lyons 1993, Elliot 1994a, 1994b). From the Australian perspective, the eventual ratification of the Protocol would constitute a major diplomatic triumph despite considerable hostility from other ATCPs. The temporary ending of the minerals controversy (the Protocol has not resolved this issue) has enabled Australia to rejoin informal information networks within the ATS and alleviated the process of resisting further criticisms of the ATS within the United Nations. Moreover, the production of the Protocol was later cited by Gareth Evans as evidence of Australia's commitment to international citizenship through the promotion of environmental protection allied with co-operation in development and human rights.

Thereafter, the Australian government under Paul Keating played an active role in promoting the legislation to enact this Protocol, and the eventual implementation of the Protocol via a Co-ordinating and Implementation Group. During the 1994 ATCP meeting in Kyoto, Australia presented proposals for implementation of the Protocol whilst at the same time opposing Chilean and French proposals for a separate annexe on Antarctic tourism. The proposal for a separate annexe on polar tourism was rejected because Australia felt that the Protocol already covered all activities including tourism and hence there was no need to develop separate regulations for the management of that particular activity. The Antarctic Division and the Department of Foreign Affairs argued, however, that the development of a liability annexe would be an important innovation in the regulation and management of scientific and tourist activities in the Antarctic. This touches upon sensitive issues such as: Who would be responsible for the consequences of an environmental accident? And how would the international community deal with environmental pollution?

In contrast to the immediate aftermath of the 1959 Washington Conference, Australian policies and public pronouncements on the Antarctic have increasingly focused on environmental protection, scientific activity and the defence of the Antarctic Treaty System. Whilst government documents continue to highlight Australia's sovereign rights in the AAT, the dramatic tones of the Cold War have been replaced by more sober reflections on territorial sovereignty and resource rights. Recent *Annual Reports* of the Antarctic Division have confirmed that there are four principal policy objectives: the enhancement of Australia's role in the ATS, the protection of the Antarctic environment, the promotion of scientific knowledge relating to the Antarctic and the administration of the AAT (DEST, Antarctic Division *Annual Report* 1992–3, Australian National Antarctic Research Expedition (ANARE) 1995). The importance of the sovereignty issue permeates ongoing policy to the extent that the Antarctic Division refers to it in order to secure funds for polar operations and the Australian delegations to the ATCP meetings use it to maintain influence within the ATS. Since the Protocol, therefore, successive Australian governments have reiterated their commitment (because of their sovereign claims) to the environmental protection of the Antarctic and Southern Ocean. The Australian claim to the AAT is increasingly regarded as being a mechanism for maintaining influence within the ATS and for the promotion of effective environmental and resource management (cf. New Zealand) in the southern polar region (Davis 1995).

Australian participation in, and susbsequent chairmanship of the Valdivia Group (Group of Temperate Southern Hemisphere Countries on Environment) between 1995–6 has been indicative of recent government interest in pursuing wider southern hemispheric environmental interests. The Australian Department of Foreign Affairs and Trade has reiterated these changing environmental and strategic priorities:

> The Valdivia Group offers the prospect of a group of countries pursuing common environmental interests across the traditional boundaries that characterise many international meetings, and in the underpinning of these common interests with credible scientific research. It has long been recognised by the scientific community that countries in the temperate zone of the southern hemisphere share much in common. All of the continents support similiar temperate environments and their climatic systems are linked through large scale oceanic and atmospheric processes. (Department of Foreign Affairs and Trade 1996)

Whilst little formal southern hemispheric environmental co-operation has been undertaken thus far, the Valdivia Group represents a major element in the rearticulation of Australia's environmental security. In alliance with Australian interests in Antarctica and the Southern Ocean, the practical geopolitical reasoning of future governments in Canberra will be judged on an ability to maintain a balance between common southern hemispheric environmental concerns and Australian resource and territorial rights in the Antarctic, the high seas and the deep seabed.

Summary

This chapter has considered the historical and political significance of Australian Antarctic involvement. The historical context for this region lies in the changing nature of Anglo-Australian relations and imperial polar policy. The declaration of an Australian Antarctic Territory (AAT) in 1933 was the culmination of nearly 15 years of negotiations. In the post-war period, successive Australian governments had striven to improve territorial and resource rights to the AAT. The policy context was shaped by the emerging Antarctic Treaty System, the Cold War and Australian relations with Britain and the United States. In the last 30 years, a significant change in policy priorities evolved as the more recent Hawke and Keating governments have stressed the importance of the environmental protection of the Antarctic as opposed to earlier concerns with territorial claims, demilitarization and economic potential of the region. This does not imply that these latter themes have diminished in importance, but it does acknowledge that Australian Antarctic priorities and the reasoning employed to legitimate them have altered from traditional concerns of territorial security and control to environmental security and resource monitoring.

5

Chile and Territorio Chileno Antárctico

Chile, at the onset of the Second World War, became the sixth nation to claim a segment of the Antarctic continent and surrounding seas. The declaration of a Chilean Antarctic Territory (CAT) had been prompted by the decision of Norway to host an International Polar Congress in 1940 at Bergen (Pinochet de la Barra 1955, 1991; Hanessian 1965). This stimulated the Chilean government to formalize their thinking on the polar continent and possible responses to rival claimant nations such as Britain and Argentina. In time, Chilean interest in the Antarctic mirrored Argentina's in the sense that both South American nations have been adamant that these claimed territories are integral parts to their national territories. As President Pinochet noted in 1975: "the government has adopted measures and traced the lines of a policy designed to preserve our solid rights in the National Antarctic territory, now incorporated as the 12th Region" (cited in Pittman 1981:1403). As with the Commonwealth states of Australia and New Zealand, Argentina and Chile have co-operated with one another over countering legal and political claims against the so-called "South American Antarctic". The major difference between these two states, however, was that Chile had been consistently more supportive of the principle of international co-operation and freedom to conduct scientific research within the CAT (Pinochet de la Barra 1995).

Chile's historical and political experiences in the Antarctic and Southern Ocean raise some pertinent themes for this account of the SORS. In conjunction with the Argentine case, Chilean involvement in these regions has been explicitly legitimated through the employment of geopolitical reasoning. Whilst this has been well appreciated, largely through the sterling endeavours of Jack Child and Howard Pittman, the connections between

Chile's involvement in the Antarctic, the international politics of the region and the role of geopolitical thinking has been insufficiently specified (eg Pittman 1981, Child 1988, Kelly and Child 1988). This account explicitly demonstrates that naturalized geopolitical reasoning continues to legitimate Chile's interests in the Antarctic and explains her later opposition to UN-led attempts in the 1980s to change the international politics of the Antarctic. As with the Australian and Argentine examples, the polar region is a distinct feature of domestic policy making in terms of budget allocations and the substantial involvement of the armed forces (Infante 1993). The role of the armed forces in supporting Chile's Antarctic programmes has frequently been remarked upon by American and British scientists and political leaders as a disturbing aspect given the commitment of the ATS to the demilitarization of the continent (Beck 1986). In spite of the restoration of democracy in 1989, the former president, General Augusto Pinochet, remains commander in chief of the armed forces (Arriagada 1986, Marcella 1990). Pinochet's vision of Chile as an Antarctic nation remains salient within the official pronouncements of Chilean politicians and diplomats. Pinochet reminded his audience at the Villa las Estrellas base in Antarctica during a visit in 1984: "You, compatriots, with the strength of spirit of pioneers are sending a light into the future from the present, which honourably represents the intention of the government and the Chilean people to develop this region" (cited in *Boletin Antarctico Chileno* 1984: 59).

This chapter is organized into six major parts: consideration is given to the historical background of the Chilean Antarctic claim and its geographical and legal foundations. Comparison with Argentina will be introduced in order to highlight the concept of a naturalized geopolitical reasoning. Chilean activities in the aftermath of the formal polar claim in 1940 are then examined with the aid of geopolitical ideas such as Terra Australis. As a claimant nation at the 1959 Antarctic Treaty, Chile's role in conjunction with Argentina is considered because she was perceived by other states to be dogmatic over the issue of national claims. The development of Chilean interests in the resource and scientific management of the Antarctic and Southern Ocean is addressed with a special focus on the Pinochet regime (1973–89). As with the late President Perón of Argentina, Augusto Pinochet's vision of Chile as an Antarctic nation was highly significant in ensuring continued participation in the ATS and regional security pacts for the Southern Ocean. However, Chile became involved in territorial conflicts with Argentina over the Beagle Channel which raised fears of possible conflict spreading towards Antarctica. The final section deals with the democratic interest in the Antarctic and the contribution of Chilean delegates in transforming some of the business of the ATS.

Chile as an Antarctic nation – geographical and geological fate?

The basis for Chile's claim to the Antarctic is analogous to the Argentine claim (Joyner 1992). In contrast to the Commonwealth SORS, Chilean claims to the Antarctic are rooted in a combination of historical, geographical and geological contexts (Pinochet de la Barra 1955, Child 1988). Present title is said to flow from the 1493 Papal *Inter Caetera*, augmented in turn by the 1494 Treaty of Tordesillas between Spain and Portugal. Chile contends, therefore, that it inherited legitimate claims to title to Antarctic lands when it gained independence from Spain in 1810 (Child 1988, Mericq 1988). Chilean claims to a geological and geographical inheritance can, it is suggested, be traced back to the 15th and 16th centuries. Moreover, Chilean authorities have also argued that their interest in *Terra Australis* was well appreciated by the hero of the Independence, Bernardo O'Higgins, who instructed vessels in the 1820s to sail south to explore the southern lands of Chile (Riesco 1985, Arnello Romo 1988a). Chile also maintains that preferential legal considerations derive from its geological continuity and geographical proximity to the Antarctic. Finally, the argument is made for effective occupation, albeit only since 1947, when the Chilean government created Antarctic bases and began to perform certain "administrative acts aimed at buttressing the legitimacy of its claim to sovereignty on the continent" (Joyner 1992:46). Previous involvement in the polar region had been limited to whaling, through the activities of the Magellanes Whaling Company base at Deception Island (Tonnessen and Johnson 1982).

This theme of geological continuity and geographical proximity recurs throughout this chapter, exploring the ways in which forms of geographical reasoning have contributed to political and legal arguments over Chilean claims to the Antarctic and parts of the Southern Ocean (Toledo and Zapater 1991). The legal fact that the Chilean polar claim of 1940 never specified a northern limit to their sectoral claim between longitude 53 and 90 West is indicative of the belief in geological continuity (see Map 5). Unofficial Chilean maps (ie those not produced by the Chilean Military Geographical Institute) or maps produced by other states have sometimes introduced a northern limit of 60 degrees parallel. This omission of a northern boundary undoubtedly reflected official Chilean thinking that there was no northern limit, since the Chilean Antarctic territory was regarded as an integral part of the Republic (Child 1988, Mericq 1988). Fernando Zegers, a Chilean polar authority and diplomat, reminded the United Nations in 1983: "By definition Chile is an Antarctic country – because of its historic mission, its geographical position, its presence there and the activities of its nationals" (UNGA Records 1983). If there was a northern limit, this had to be located closer to Bolivia and Peru rather than somewhere in the Drake's Passage or the South Pacific (Riesco 1985, Arnello Rono 1988b).

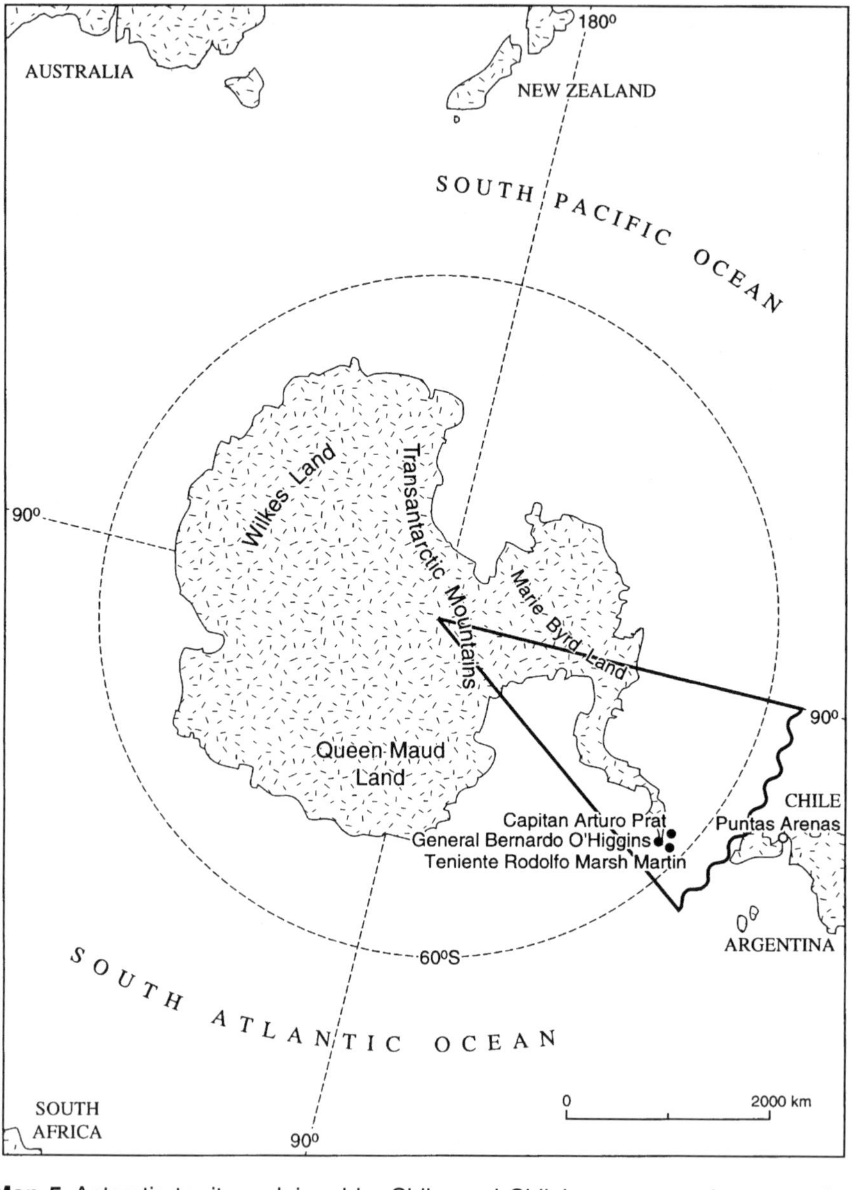

Map 5 Antarctic territory claimed by Chile, and Chile's permanent bases on the continent

The deployment of geographical proximity to legitimate the Chilean Antarctic sector drew substantially on earlier Canadian arguments about a polar sector stretching towards the North Pole. The British government applied so-called "sector theory" to the Antarctic and declared a Falklands

Islands Dependency. There remain many other legal uncertainties over the sector theory, even though Argentina and Chile have argued that their geographical proximity to the continent entitles them to special rights of possession (Joyner 1992:60). International lawyers outside South America have maintained that geographical proximity is a poor basis for a legal claim: first, the definition of proximity is stretched given that there are 600 miles between the Antarctic and the South American mainland; and second, these arguments are usually intended for resolving conflicts between coastal states and disputed offshore islands. In short, Christopher Joyner contends that: "In sum, there is overwhelming concurrence that polar sectorization through propinquity serves primarily as a political convenience for particular parties . . . Propinquity between Antarctica and the other continents is more fiction than fact" (Joyner 1992:61).

This argument over geographical proximity or even geological continuity (an argument favoured more by Argentina than Chile) is important not because of its dubious legal merit but rather for its contribution to a long-standing naturalized geopolitical discourse which has argued that Chile's national boundaries as such are determined by natural geographical features such as mountain chains or river basins rather than by social or political criteria. These arguments were deployed in the 19th century when Chile was negotiating its Andean boundaries with Argentina and resolving its northern boundaries in the aftermath of the War of the Pacific (Morris 1995, 1996). The Chilean claim to the Antarctic reflects this longer-standing context of the international politics of South America and the accompanying territorial uncertainties. Subsequent Chilean geopolitical thinking has been strongly influenced by these naturalized geographical discourses, and the unshakeable belief remains that Chile "lost" territory in the Andes, Atacama desert and Tierra del Fuego regions (Barton 1997).

The shift of Chilean geopolitical thinking in shaping Antarctic perspectives has been reviewed (Child 1988:105–32). Geopolitical reasoning has been employed by government bodies and influential individuals such as President Pinochet in order to raise public awareness of Chile's Antarctic claims. Chile's Antarctic activities are organized by major institutions such as the Instituto Naçional Antarctico de Chile (INAC) and the military staff attached there support these programmes (Bustamente 1988). Geopolitical thinking is a major topic of Chilean war academies and a regular feature in military journals such as *Revista De Marina*. The interaction between civilian and military institutions and government departments is a substantial aspect of South American polar programmes. Geopolitical ideas have also provided a strategic rationale for Chilean claims that her interests in the Antarctic are linked to maritime concerns in the South Pacific and the South Atlantic. As the Chilean geopolitical thinker Riesco (1985:26) noted in Mackinderian terms: "Whosoever controls the waters of the South Pacific Ocean projects geopolitical control over the west coast of South

America. Whosoever controls the South Pacific Ocean simultaneously exercises control over the Antarctic continent."

From the 1940s onwards, it was proposed that a Mar de Chile stretched from Chile's northern borders across the South Pacific towards the Isla de Pascua (Easter Island) and then southwards to the Antarctic continent. This cartographic and legal invention had two functions: it enabled Chilean military leaders to argue that Chile was an organic territorial entity with distinct natural boundaries and it would justify subsequent claims to an exclusive economic zone. Under Supreme Decree 346, dated June 1974, the Pinochet regime officially declared a Mar de Chile but without specifying the exact geographical limits (Glassner 1985:335, Child 1988:116). Geopolitical ideas such as Chile possessing a Mar de Chile have also contributed to raising the profile of such features in education and popular culture. Under this kind of argument, which has strong parallels with Argentina, the Chilean Sea links together the constitutent parts of Chile: mainland, Antarctic and insular (Easter Island, Beagle Channel islands, Juan Fernandez Islands et al). Overall, Chile's claim to these southern and western territories is superior to that of other nations including Argentina because as Canas Montalva noted in 1948: "To be the most southerly country on earth and therefore the most Antarctic, will permit her [Chile] an invaluable continuity of her territory between the American continent and Antarctica" (cited in Pittman 1981:1032).

Geopolitics of Terra Australis: Chile in Antarctica

Following a report commissioned by Professor Escudero Guzman in 1939, the Chilean government of President Pedro Aguirre Cerda demarcated a Chilean Antarctic Territory. In November 1940, therefore, Presidential Decree 1747 (cited in Pittman 1981) dramatically increased the geographical area of the state of Chile:

> Considering that it is the duty of the state to fix its territorial boundaries with exactitude . . . The Chilean Antarctic or Chilean Antarctic Territory is formed by all lands, islands, islets, reefs, glaciers and pack ice, and others, known or unknown, and the respective territorial sea existing within the segment constituted by the meridians 53 degrees longitude west of Greenwich and 90 degrees longitude west of Greenwich.

The Chilean Antarctic Territory was subsequently placed under the provincial jurisdiction of "Magellan and Chilean Antarctica" with a capital based in Puerto Williams.

On the eve of the Second World War, five states had pressed formal claims to the polar continent: Britain, France, Australia, New Zealand and Norway. Three-quarters of the Antarctic had been claimed by these states and pressure was growing within Chile to react to these political developments. In his position as Chilean chief of the southern military region Canas Montalva was particularly sensitive to the strategic importance of Antarctica and the shipping routes of the South Atlantic. He therefore suggested in 1940 that Chile had to establish a territorial claim to the region because:

> Chile must call attention to herself, beside the other powers that claim ownership over the southern polar hemisphere, precisely fixing the logical boundaries of her proprietorship, even though her modest resources have not permitted this until now, establishing the sovereignty that was necessary or giving it the commercial exploitation that it offers. (cited in Pittman 1981:1189)

The Cerda government was concerned that American expeditions were preparing the legal ground for a possible US territorial claim (Pinochet de la Barra 1955, Hall 1986). In 1939–40 the Roosevelt government had informed the Chilean and Argentine governments that American activities in the Marie Byrd Land and Ellsworth Highlands was predicated on the assumption that the western part of the Antarctic should be controlled by the "American" republics in order to maintain regional security and freedom of movement around the Antarctic peninsula (Klotz 1990). Whilst the American government urged Argentina and Chile to participate in polar activities, their claims to the Antarctic region already overlapped with Britain (Pittman 1981).

The outbreak of war in Europe and Asia between 1939 and 1941 meant that northern hemispheric claimant states were preoccupied with other matters. In March 1941, the Chilean government hosted a meeting in Santiago with their Argentine counterparts in order to discuss the articulation of a common polar boundary (Labra 1947). The subsequent Escudero–Moreno Accord failed to establish such a barrier but did conclude that a geographical region called the "South American Antarctic" existed and that only Argentina and Chile enjoyed rights to this polar space. In the official statement covering the Escudero–Moreno Accord, geographical factors feature strongly in the articulation of a Chilean polar sector:

> The Chilean Antarctic has also undeniable strategic value. Graham Land and Tierra del Fuego form two headlands which are the true guardians, if we may call them as such, of the Drake strait . . . There would exist on our part the obligation to preserve and defend it precisely for this [the sea lanes around Cape Horn], because it is the southern confine of Chile, an integral part of its soil.

The arguments concerning geographical disposition and geological continuity were combined to produce a context from which Chile's strategic responsibilities were articulated by the Cerda government. Geographical fate, therefore, was considered compelling and Chile's attitude towards the Second World War was deemed less problematic to the Allies compared to Argentina given its declaration of support to the USA. One enduring irony, however, was the usage of the term "Graham Land" (a British term) to describe the Antarctic peninsula which had already been claimed by the British in 1908 and 1917.

The two South American governments later agreed that substantial efforts should not only be made to occupy the Antarctic region but also to remove the colonial presence of Britain (Labra 1947, Montalva 1948). The first stage was the creation of a polar commission in 1942 to co-ordinate Chilean Antarctic expeditions. The second stage of this process was carried out by the Military Geographical Institute (MGI) with the production of new Antarctic maps which transformed a host of geographical features such as the Antarctic peninsula into the Peninsula O'Higgins (see Montalva 1948, Pinochet de la Barra 1976).

After the Second World War, the Chilean government announced that an expedition would be launched in January 1947 (Labra 1947). The first permanent Chilean base (named Arturo Pratt) was constructed later that year and located on Greenwich Island, South Shetlands. In a speech made by the Chilean Minister of Foreign Affairs Raul Gomez, it was announced (with direct parallels to the Moreno accord) to the Chilean Senate that:

> The Chilean polar sector is the natural prolongation of our continental territory. If to this we add that Chile is the nearest country in the world to the Antarctic – and I have here the other geographical fact in our favour, that of contiguity or neighbourhood – we can categorically state that titles of geographical character give the Republic preferential rights of sovereignty over the polar zone delimited by the supreme decree of 1940 . . . The Chilean Antarctic has also undeniable strategic value. Graham Land and Tierra del Fuego form two headlands which are true guardians, if we may call them such, of the Drake strait . . . Apart from this, when in the future lines for transpolar aerial navigation are established, as is planned, they will give Graham Land a preponderant role in international traffic . . . There would exist on our part the obligation to preserve and defend it . . . because it is the southern confine of Chile, an integral part of our soil. (Cited in Pittman 1981:1032)

The MGI sent six military officers with the first expedition to the Antarctic and provided expertise in producing public maps and charts depicting Chile as an Antarctic nation. The President of Chile, Gonzalez Videla, later visited the South Shetlands in February 1948 to formally inaugurate a new station (Bernardo O'Higgins) in Trinity Peninsula (Hanessian 1965:15). He announced to the Chilean public that: "To all in Chile, which I am certain is dwelling on this memorable act, I award this land of tomorrow, sure that its people will know how to maintain energetically the sovereignty and the unity of our territory from Africa to the South Pole". The responsibility for the Chilean Antarctic programme rested with the Chilean navy under the Naval Ministry's Antarctic Section and the Chilean Antarctic Commission under the Foreign Ministry (see Mericq 1988). In 1948, an Antarctic section was established by the General Staff of the Armed Forces and invested with the duty of co-ordinating and directing military research into the Chilean Antarctic Territory (CAT).

For some outside observers, the Chilean Antarctic programme was motivated by the changing political situation in the polar region. The Canadian ambassador to Chile Fraser Elliot noted to the Department of External Affairs in October 1948 that: "There seems little doubt that Chile's decision was taken in anticipation of the forthcoming international discussions of the Antarctic question and in answer to Britain's determination to establish another base in the Alexander Land area." The international politics of the Antarctic were undoubtedly influenced as other states such as Britain established national programmes and/or launched high-profile expeditions as did the USA. The British had already established claims to the Falkland Islands Dependencies and this overlapped with the new Chilean claim. The United States navy had established an Antarctic Development Project with an *Operation High Jump* and *Operation Windmill.* Both operations were designed to improve polar training and consolidate US interests in the Antarctic (Beck 1986, Hall 1986, Klotz 1990). The USA had also established an inter-American security system under the Rio Treaty of September 1947 (Child 1985). Whilst the American administration was more concerned with the principle of hemispheric collective security, South American states such as Argentina and Chile argued that the principles of the Treaty could be applied to the Antarctic. For example, when the British deployed warships such as HMS *Nigeria* in 1948 in response to South American "intruders", Chile argued that under the terms of the Rio Treaty that constituted an "armed threat" and should be dealt with by the Inter-American security system. The Truman administration rejected this request during the 9th International Conference of American States in May 1948 and instead proposed that a study group should be created to consider the territorial and political status of Antarctica (Dodds 1996c).

In June 1948, the Americans introduced their first proposals to internationalize the Antarctic, either through trusteeship or condominium so that arguments over territorial claims could be resolved (Hall 1986). Chile rejected these earliest proposals and refused to negotiate on the grounds of national sovereignty. In their formal reply to the United States Embassy in Santiago, the Chilean minister of foreign affairs noted that "It was impossible for the Government of Chile to accept the ideas suggested by the Department of State . . . a government such as that of Chile which exercises full and effective sovereignty over its Antarctic territories would be placed on a footing of complete equality with those governments which have only claims to make" (National Archives of Canada, Department of External Affairs, 1948).

The formal reply to the American proposal was also indicative because Chile reminded the Truman government that the Rio Treaty had declared a part of the Antarctic (between 24 degrees West and 90 degrees West) "a security zone of the Americas". The geopolitical significance of this argument was that the British presence in the Dependencies of the Falkland Islands constituted a threat to the territorial integrity of the Argentine and Chilean Republics. If this point had been accepted then the American community would have had a legal and political duty to remove these "threatening" items from the American continent, but the United States had not envisaged Britain as a threat given its fears of communism spreading throughout Latin America. In spite of these differences, Chile did negotiate in an important naval agreement with Britain and Argentina which restricted the movement of warships, throughout the 1950s, in Antarctic waters fearing a possible confrontation (Dodds 1996c). This was subsequently renewed annually until the Antarctic Treaty effectively demilitarized the region and henceforth superseded that Naval Agreement.

Chile's initial hostility to the American proposals to internationalize the Antarctic was followed by a diplomatic initiative led by the principal Chilean Antarctic representative, Julio Escudero Guzzman (Pinochet de la Barra 1995:8). In negotiations with the US envoy Casper Green, Escudero proposed an agreement whereby all claims and rights to the Antarctic would be effectively "frozen" and where scientific investigation in the region would be encouraged. It was further proposed that the seven claimant states and the United States would form a consultative committee in order to regulate the exchange of scientific information. From the Chilean point of view, the Escudero agreement was attractive because it avoided any suggestion of an international authority intervening in the management of the Antarctic; it excluded other parties such as the Soviet Union and it protected Chilean claims in the face of Argentine and British activities and counter-claims. In September 1949, the American delegation handed a copy of the Escudero proposal to the British government for initial reactions. The Escudero proposal was effectively shelved because the following year the Soviet Union issued a declaration to all claimant states informing them that the USSR would have to be a participant in any new international regime for the Antarctic (Hall 1986). In spite of their political differences, the United States and its Western allies were anxious to avoid the participation of the Soviet Union in any international polar forum (Toma 1956). The onset of the Korean War (between 1950 and 1953) later postponed active interest by the superpowers in the international politics of the Antarctic and allowed South American states further to consolidate their territorial claims.

Whilst the international politics of Antarctic were undergoing transformation in the late 1940s, many of the earliest texts on Chile's interest in the Antarctic were published in a journal called *Revista Geografia de Chile – Terra Australis* by the National Committee for Geography, Geodesics and Geophysics between 1948 and 1971. In particular, the seminal figures of Canas Montalva and Pedro Ihl were responsible for constructing propositions about Chile and the Antarctic (Pittman 1981). They were notable in promoting a geopolitical consciousness devoted to stressing the emotional and geopolitical aspects of the Antarctic, the Southern Ocean and the southern parts of Chile. Within the pages of *Terra Australis*, Canas Montalva expounded the view that the construction of a Chilean base in Antarctica would epitomize a new era for Chile and the Antarctic region. In future, he argued that Chile, as a South Pacific and Antarctic nation, would have to undertake the geopolitical responsibility of continental defence and the protection of her Antarctic claims. The recent signing of the 1947 Inter-American Treaty of Reciprocal Assistance (the Rio Treaty) seemed to confirm Canas Montalva's arguments that Chile had a responsibility for defending American interests in the South Pacific and around the Antarctic peninsula (Montalva 1948). The defence of the Chilean Antarctic claim could, therefore, then be located within a wider context of hemispheric responsibility (Ihl 1952).

Throughout the 1950s, various Chilean governments sought to maintain an Antarctic programme encouraging base construction, resource exploration and surveying. The period was dominated by political and legal tensions with Argentina and Britain. Chilean bases or huts in the Antarctic were being rapidly constructed in response to Argentine and British occupation of the Antarctic peninsula in the light of British attempts to subject these rival polar claims to the scrutiny of the International Court of Justice (ICJ) between 1947 and 1955 (Dodds 1996c). In response to the British application, the Chilean Ministry of Foreign Affairs declared that the CAT was an internal Chilean matter and therefore not subject to the jurisdiction of the ICJ (Pinochet de la Barra 1955). Moreover, Chilean governments attempted to draw the United States into supporting South American claims to the Antarctic on the basis of the provisions of the 1947 Rio Treaty. By 1952, Chile had entered a military assistance pact with the United States, and the Chilean armed forces were anxious to obtain military assistance from Washington. As Chile was not considered by the United States as a leading player within the Inter-American Security System, it was unlikely that the USA would have intervened in the manner hoped for by the government in Santiago.

Prior to the Antarctic Conference in October 1959, successive Chilean governments had expressed widespread opposition to plans for a change to the status of national claims to the Antarctic. The objections of the Alessandri government (1958–64), for example, were based on two related points. First, Chilean Antarctic Territory was considered part of the Republic of Chile and its political and legal status was not negotiable. Hence, Indian proposals to internationalize the Antarctic in 1956 through the involvement of the United Nations were rejected. The Chilean foreign minister, Enrique Gajardo, had issued a communiqué rejecting these plans: "The Chilean government considers that this matter, relating to territories over which Chile has sovereign rights, should not be brought to the attention of the world organization [ie the United Nations] for debate" (cited in *El Mecurio* 23 February 1956). The Latin American caucus within the UN played a key role in forcing the Indian delegation to withdraw this proposal in September 1956. Second, the Antarctic peninsula was considered part of the South American continent and thus was of little concern to those outside the American continent. In a statement by the foreign minister, Alberto Sepulveda, on 18 February 1958, these two themes are effectively joined together:

> There is an American Antarctica, which is an integral part of the western hemisphere . . . The Antarctic territory of Chile is a part of the American security zone . . . Our country holds the oldest rights of sovereignty on this territory, as established in the first place by repeated provisions and mandates from Spain and then later, though our life as a Republic . . . The Government of Chile could only reject any proposal which might imply the internationalisation of, or condominium over, any part of its Antarctic territory, whether it be in Antarctica, in America or in its insular possessions in the South Pacific. (National Archives of Canada, Department of External Affairs 1958)

Chilean attitudes over territorial claims had already hardened before the scientific programmes of the International Geophysical Year (IGY). At the first meeting of the CSAGI in July 1955, convened to discuss the creation of the IGY, the Chilean delegation headed by their French ambassador argued that the proposed location of bases in the Antarctic peninsula was being decided on political rather than scientific grounds. In conjunction with their Argentine colleagues, it was suggested that the Committee should pass a resolution which agreed that "these resolutions [for the IGY] do not modify the existing status in the Antarctic regarding the relations of participating countries" (cited in Hall 1986:33). Whilst there was general sympathy for further scientific co-operation and the establishment of scientific bases, the legal and political protection of Chilean Antarctic Territory was considered paramount in the face of Soviet proposals to create a base in the Antarctic peninsula (Hall 1986). The CSAGI later accepted a Chilean proposal that explicitly noted that the scientific programmes of the IGY would not be considered to have any bearing on existing territorial sovereignties in the Antarctic and Southern Ocean. The "gentleman's agreement" negotiated in Paris accepted that participating nations would not engage in "legal and political augmentation" during the period of the IGY (Beck 1986:48).

This informal suspension of the sovereignty disputes in the Antarctic should not be over-estimated, however. The Chileans were engaged in numerous diplomatic protests with their British and Argentine counterparts over the production of maps depicting national polar claims and the establishment of scientific bases in particular geographical sectors. During the IGY, for example, the three claimant nations re-negotiated the 1948 Naval Agreement and attempted to avoid a return to the confrontational episodes of the early 1950s (Dodds 1996c). The most pressing concern for

Chile, however, was the prospect of other nations such as the Soviet Union occupying bases in the CAT. In their discussions with other claimant states, such as those held with Australia and Argentina between 1956 and 1958, Chilean representatives expressed concern over Soviet activities and confirmed that they were anxious to improve co-operation in the light of their joint concerns. The decision by the US government in July 1957 to propose either an extension to the IGY or the creation of an international commission for the co-ordination of science was not welcomed by the Chilean government. The latter, in conjunction with Argentina and Australia, wanted reassurance from the Eisenhower government that the Soviet Union would not be encouraged to remain in the polar region (Klotz 1990). The Americans were neither willing to give that political guarantee nor were they prepared to recognize the formal claims of these states.

In May 1958, the US Department of State issued an invitation to the 11 other participants in the IGY to consider creating a new international regime for the Antarctic. During June 1958, the 12 parties convened in Washington in order to consider the political fate of the Antarctic. In their formal reply to the American offer, the Chilean government reaffirmed that their territorial claims were an integral part of the nation and that the scientific activities of the IGY should not be used to press or improve upon any claims to territory. The Chilean government had participated in the preliminary meetings because it was widely appreciated that the American administration was intent on creating an international regime which would neutralize the Antarctic region from the Cold War. The United States was anxious that Soviet participation in the Antarctic should be restrained through agreement on demilitarization and scientific freedom and rejected appeals from southern hemispheric nations that the Soviets should be excluded from these negotiations. The Chileans were also aware that the United States was not going to recognize their claim to the Chilean Antarctic Territory at the expense of compromising relations with Britain and to a lesser extent with Argentina. Moreover, the American government had announced that it reserved all rights to the Antarctic region and did not preclude the possibility of a future claim. Further, the Chilean government wanted to avoid the possible interference of the United Nations given the earlier Indian proposals in 1956 and 1958. In the light of these uncertainties Chile was anxious for new discussions. As with their Argentine counterparts, participation in the subsequent Washington Conference was based on an assessment of the political, strategic and legal situation (Pinochet de la Barra 1991).

The eventual Washington Conference between October and December 1959 was the product of over 60 preparatory meetings concerning an agreed approach to the Antarctic. The Chilean delegation was determined that no undermining of Chilean national sovereignty in the CAT would be accepted. The result of this position meant that Chile was hostile to initial proposals for unlimited access to the Antarctic by other interested parties;

the compulsory jurisdiction of the International Court of Justice in the event of dispute with another party; the non-specification of a time limit for any new international regime and the creation of a permanent secretariat and the application of regulations concerning criminal jurisdiction. Chile was, however, a strong supporter of early Argentine proposals for the inclusion of the high seas within any zone of application and the complete prohibition of military activities within the region because the Alessandri government did not wish the Antarctic to be transformed into a militarized space. The issue of the zone of application and non-militarization had been problematic as the British and American delegations argued that the provisions of an Antarctic treaty should not apply to the high seas. The progress of these preliminary meetings was not disrupted by disagreements over the zone of application and the Chilean delegation under Marcial Mora attended the Washington Conference.

Chile and the 1959 Antarctic Treaty

In his opening declaration to the Washington Conference, the head of the Chilean delegation, Marcial Mora, reasserted his country's official position:

> This interest in the Antarctic Continent on the part of Chile is certainly not something of today or yesterday. It goes back to the days, now far remote, when shortly after the Republic was established the country was able, to the extent of its capabilities to broaden and strengthen its activities of all kinds in those regions . . . From a geographical point of view I wish to emphasise a well known fact: my country is the closest one to the Antarctic continent, for between the Antarctic continent and its southern most insular possessions in the Americas, the distance, is hardly more than 428 nautical miles, clear proof that the southern extreme of the American hemisphere and the Antarctic continent of today were joined together in another age. (Scott Polar Research Institute Archives 1959)

This opening speech was identified by reporting American newspapers as one of the most militant over the issue of national sovereignty. The *Washington Post* noted that three nations – Chile, Australia and France – appeared the most dogmatic over territorial claims (15 October 1959).

Subsequently, the Chilean delegation was generally regarded by Western newspapers and governments to have been supportive of the general political agenda rather than diplomatically intransigent. Some issues clearly merited considerable diplomatic interest such as the Argentine proposal for a complete ban on peaceful nuclear explosions in the Antarctic. This was supported by Chile along with New Zealand and South Africa. Argentine reservations over freedom of scientific investigation were supported in part by Chile, who was concerned that national sovereignty should not be compromised. A significant initial intervention by the Chilean delegation appears to have been the demand that reference be made to the 1947 Rio Treaty within the preamble of the 1959 Antarctic Treaty. From the Chilean perspective, the inclusion of the Rio principles would have been a remarkable reaffirmation of the American Antarctic and the responsibilities of the United States to the Southern Cone. Ambassador Mora's speeches to the Washington Conference emphasized that geopolitical linkage of the Antarctic and the American continent. The Chilean demand was later rejected by the Commonwealth delegations who insisted that if the Rio Treaty was cited then other regional security organizations such as SEATO and ANZUS should also be listed. The Chilean delegation withdrew their request in November 1959.

The eventual ratification of the Antarctic Treaty by the Chilean parliament was based on pragmatic political and security grounds. Article IV of the Antarctic Treaty, whilst not formally recognizing Chile's territorial claims to the CAT, had not undermined it. Article V had banned the usage of nuclear weapons or testing from the polar region and Article II had secured freedom of scientific investigation without compromising national sovereignties. Initial Argentine and Chilean objections to Articles II and IV were overcome after some powerful lobbying by the US and British delegations (Pinochet de la Barra 1991). However, the acceptance of Article V by the Washington delegates was the impetus for persuading the Chilean Parliament to approve domestic legislation in March 1961 incorporating Chile's acceptance of the provisions of the Treaty.

With the demilitarization of the Antarctic continent accomplished, successive Chilean governments whilst maintaining a substantial presence in the CAT have concentrated on promoting the territorial and economic integration of the state. As with the Argentines, Chilean governments have argued that Article IV only suspended territorial claims to the Antarctic rather than removing them from the political map of the Antarctic. Indeed as Oscar Pinochet de la Barra later recalled:

> Senator Mora convincingly asserted that the permanent 'freezing' of Chilean sovereignty would be tantamount to renouncing it. Some days later, a proposal was made to the effect that the Treaty should not be completely impervious to change and be capable of modification . . . its revision would be allowed by a majority vote after it had been in force for 30 years. This was regarded as a provision whose application would obviously not be mandatory. (Pinochet de la Barra 1991:207)

The Chilean delegation had been successful in insisting that sovereign rights to the Antarctic continent could not be suspended forever and instead there had to be a degree of political and legal flexibility.

Resources, regional interests and the Antarctic region

Chile, it has been argued, is the most maritime conscious of all the Latin American nations (Morris 1989b, Barton 1997). This geopolitical disposition is usually considered a product of Chile's historical evolution, geographical shape and political relations with neighbouring Argentina (Pittman 1988b:176). The development of geopolitical concepts such as the Chilean Sea (Mar Chileno) and the Arc of the Southern Antilles in the 1950s have illustrated this concern for resource exploitation, polar geopolitics and maritime control.

The Chilean government established a substantial research programme through the institutions of the Instituto Antarctico Chileno, the armed forces and the Antarctic Commission (Pinochet de la Barra 1976). These scientific projects have been supplemented by an array of administrative acts and activities which have ensured that the "effective occupation" of the CAT has been a persistent policy theme. Whilst the legal and political protection of the CAT was linked to Chile's regional interests in the South Pacific, the Frei government (1964–70) was not involved with wider geopolitical proposals such as an Argentine plan to create a South Atlantic pact in the 1960s or 1970s. Politically speaking, the Frei and Allende governments refused to follow the diplomatic line of the military regimes in Argentina to isolate post-revolution Cuba or to reduce trading relations with the USSR and Eastern European nations.

The most pressing issues for Chilean governments in the 1960s and 1970s revolved around the legal and political protection of territorial sovereignty, the avoidance of armed conflict in the region and the regulation of resources within the CAT. This became more evident under the Pinochet regime between 1973 and 1989 when Chile had been an active participant in scientific research into living resources. The creation of the BIOMASS programme in 1976 was designed to further understanding of the structure and dynamics of the Antarctic marine ecosystem as the basis for the future management of potential living resources. Chile also organized oceanographic and resource surveys of the Pacific sector of the Southern Ocean and, in conjunction with the Argentines and Australian participants, has consistently reminded other nations that these living resources raise fundamental questions over maritime jurisdiction and sovereign rights (Beck 1986:102–3). That the latter had been a persistent feature of Chilean contributions during the earliest meetings of the ATCMs is shown by their objection to a British proposal to involve outside bodies such as the International Court of Justice. The chairman of the Chilean delegation to the 1979 ATCP meeting in Washington also conveyed these long-standing concerns of his country:

> The Chilean government considered it imperative to design parameters for activities that will lead to the discovery of new resources within the solid framework of our Antarctic regime . . . To solve this problem we must begin negotiations by dealing with the possibility of establishing a system that will enable the Consultative Parties of the Antarctic Treaty to establish regulatory standards of procedure for the exploration and exploitation of the

> aforementioned resources . . . Our situation as the country nearest to the Antarctic places us in an advantageous position at the same time that it makes us very vulnerable to any changes in the Antarctic environment caused by uncontrolled activities. (ATCP 1979)

Throughout the resource-based negotiations over the CCAMLR and CRAMRA in the 1970s and 1980s, Chile's position hinged on the following three points: (1) support for the development of resource regimes and the eventual exploitation of these resources if the rights of claimant states were upheld; (2) the conservation of the polar environment would be a policy priority given Chile's geographical proximity; (3) the ATS would have to be sensitive to the repercussions of the ongoing Law of the Sea negotiations relating to the rights of coastal states and the rights of the international community expressed in terms of a common heritage of mankind. Fernando Zegers, for example, warned in 1979–80 that the ATS would have to deal with these issues and later proposed a southern hemispheric conference in order to discuss *inter alia* non-living resource proposals. Zegers later consulted again with New Zealand and Argentine colleagues at the Law of the Sea negotiations in 1982 (Zegers 1983).

During the CCAMLR negotiations in the late 1970s, Chilean delegations argued that the regulation of living resources such as krill was vital given existing claims over maritime jurisdictions and the fishing activities of Japan, Russia and the Republic of Korea. Earlier preoccupation over maritime jurisdiction is evident in Chilean statements released after the negotiations for the Convention for the Conservation of Antarctic Seas in 1972. The Chilean delegation concluded that the Convention would not: "confirm, deny or impair the rights of the Contracting Parties as regards their maritime jurisdictions and their declared juridical position on this matter" (*Handbook* of the ATS 1990:4104). In 1981, the total catch of krill in the Southern Ocean was recorded as 448 000 tons which represented a massive increase in the figures recorded for the early 1970s (Chaturvedi 1996:30). As a state neither self-sufficient in hydrocarbons nor food resources, the Chilean delegations were anxious to develop resource regimes dedicated to the conservation and management of fish stocks due to competition from technologically advanced nations such as Japan.

The ongoing negotiations over a minerals regime (CRAMRA) were also supported by successive Chilean delegations. The geopolitical rationale for this position was similiar to earlier statements on the CCAMLR negotiations. Chilean commentators attached to bodies such as the Antarctic Commission were worried that the CCAMLR negotiations had set an important precedent of non-ATCP members such as Spain participating in ATS decision making. Along with other southern hemispheric nations such as Australia, Chile was anxious to protect its sovereign interests in the face of the intentions of the United States, which had proposed that access to

Antarctic non-living resources should be open to all regardless of prior claims. Chilean scientists and politicians were also concerned that the future stability of the Antarctic Treaty System might be compromised in the haste of some ATCPs to develop a minerals regime due to fears of outside interference from the Group 77 or environmental movements. As the leader of the Chilean delegation, Fernando Zegers, noted at the Buenos Aires ATCP meeting in October 1981: "The Antarctic Treaty and the System to which it has given rise should be preserved, with all the characteristics and expressions described earlier . . . In this regard, it is essential that the sovereign rights contemplated and protected by Article IV of the Antarctic Treaty be respected and held valid, considering the accommodation that may be necessary of its interests and those of mankind" (ATCP 1981). In conjunction with other ATCPs, Chile's line had softened in comparison with earlier statements about the inalienable sovereign rights of claimant states.

Throughout the six years of the CRAMRA negotiations, the Chilean and Argentine delegates remained the most hawkish. Every opening address to the ATCP meetings was also accompanied by a reminder that Chile was an Antarctic nation with territorial rights to the CAT and further re-enforced by Chile's submission to the 1984 secretary-general's *Report* on the "Question of Antarctica". This report reiterated a range of legal, geographical and historical reasons considered indicative of Chile's sovereign interests in the Antarctic and Southern Ocean. Most interestingly, the submission concluded with a section entitled "the dangers of new Utopias":

> Chile sees the "Question of Antarctica" as a matter of taking advantage of an existing mechanism which over a period of time has provided solid evidence of its reliability and positive results. It is neither logical nor appropriate to propose new projects of which no one can foresee the outcome . . . There is no room for theoretical exercises or utopian speculation . . . The greatest merit of the Treaty is that it succeeded in bringing peace and security to a region which is fraught with tensions and which, because of its strategic position, could easily have become an arsenal threatening the world. (UNGA 1984 Part II:41)

The Chilean report did not address alternatives to the existing territorial status of Antarctica and the political system that manages the polar continent.

Two new major Chilean priorities emerged at the end of the 1980s and were linked to improving the effectiveness of the ATS and developing environmental protection measures for the Antarctic. The first theme was closely related to the "Question of Antarctica" and the challenge of developing nations to the political authority of the ATS. Chilean authorities such as Pinochet de la Barra, Zegers and Vicuna have reiterated their government's belief that the mechanisms of the ATS should not be undermined or replaced by a UN Committee for the global management of the Antarctic. In that vein, the Chilean delegation later proposed *inter alia* that information derived from ATCP meetings should, where possible, be distributed to interested parties such as the UN in order to counter accusations from states such as Malaysia and Zaire that the ATS was secretive.

The report that the Argentine ship *Bahia Paraiso* had sunk, off the Antarctic Peninsula in the South Shetlands in January 1989 made an impact on the Chilean delegation charged with negotiating their national interests inside the ATS. The effect of this incident on South American mindsets has been considered by outside commentators: "There is a growing appreciation that any environmental disaster in Antarctica will exert its first external effects upon the countries closest to the frozen continent, and hence the much vaunted proximity of Argentina and Chile becomes a liability instead of an asset". Chilean attitudes towards conservation, national sovereignty and resource exploration became increasingly ambivalent, whilst it was recognized that Chilean environmental interests would have to be protected in the event of mineral exploitation. During the preparatory meetings for the 1989 ATCP meeting in Paris, the Chilean delegation submitted a working paper which noted that:

> Chile is concerned about the danger of damage to the Antarctic environment and its inter-dependent associated systems by reason of the significant increase in human activity in the area . . . In the light of the above, the Delegation of Chile will be requesting inclusion as a top priority item on the agenda of the 15th Consultative Meeting on the subject of "Overall Measures for the Protection of the Antarctic environment and its related or dependent ecosystems against the effects of human activity" and will put forward a concrete proposal on the subject . . . We believe the time has come for an overall approach to the measures intended to protect ecosystems against human activities. (ATCP 1989)

At the 15th ATCP meeting in Paris, the Chilean delegation reiterated its support for a new set of measures dealing with the protection of the Antarctic environment. The major event of that meeting was, of course, the Australian–French proposal for a new Protocol on Environmental Protection rather than ratifying the CRAMRA agreement. Latin American countries such as Argentina, Chile and Brazil were ambivalent in contrast to those who were either openly opposed (as were Sweden, Italy) or were supportive of CRAMRA (as were the UK, the USA, Norway) during the Paris meeting. The Chilean delegation had urged the creation of an agreement which would be similar to the Agreed Measures of 1964 concerning the protection of flora and fauna and was concerned that the Australian–French proposal would involve months of further negotiation and accommodation. These misgivings were also apparent at a special meeting of the ATCPs in Vina del Mar in December 1990 when the Chilean delegation distanced itself from the draft protocol circulated by the UK, the USA, Argentina, Uruguay and Norway which contained details of the provisions and principles required for effective environmental management. The Chilean delegation whilst reiterating its support for the development of effective environmental measures was hostile to the creation of new institutions to manage these provisions which might have led to the creation of a permanent regulatory institution at the expense of the individual rights of sovereign states.

The head of the Chilean Delegation, Oscar Pinochet de la Barra, later recalled at the 1991 Bonn ATCP meeting the Chilean motivation for the eventual signing of the Protocol:

> At the end of thirty years, we were sorely tempted by the mineral exploitation issue, yet we found a way to emerge victorious and united from trial. The temptation was difficult to withstand, since reconciling science and business, peace and economic performance in the remote world of the Antarctic seemed so attractive a proposition . . . Now that the pressures to engage in the economic exploitation of Antarctica have been taken off for at least the next fifty years, there is a risk of our going to the opposite extreme and of taking over meticulous care over ensuring its purity that it will become cut off from mankind. . . . What we are seeking is an Antarctic continent that is uncontaminated but is useful to mankind. (ATCP 1991)

The Chilean position throughout the CRAMRA/Protocol negotiations appears to have been based on practical political considerations. Zegers had already noted in 1987 that the Chilean delegation would not fight for CRAMRA if it meant that the viability of the ATS was compromised (ASOC 1987). In the final analysis, the Chilean delegation in Madrid signed an agreement which did not compromise its sovereign rights over the CAT; protected the decision-making authority of the ATS; created conventions and measures for the protection of the Antarctic environment; provided for the continued political exclusion of the UN, the EU and ecological movements from mainstream decision making and ensured that the mineral wealth of the Antarctic peninsula would be left untouched until at least 2041. Pinochet de la Barra concluded in 1995: "Now we recognise that Antarctica is a natural reserve – not an international reserve made by man, but a reserve made by nature. Antarctica is dedicated to peace and science and all activities related to mineral resources are prohibited for at least fifty years" (Pinochet de la Barra 1995:10).

President Pinochet and the Chilean Antarctic Territory

The overthrow of Salvador Allende's democratic government in 1973 ushered in the military regime of Augusto Pinochet. As a former professor of geopolitics and author of *Geopolitica*, Pinochet has consistently argued that Chile's Antarctic Territory is an integral part of the Republic (Pinochet 1984). President Pinochet's 17-year domination of Chilean political life encompassed many important events regarding the Antarctic and Southern Ocean. The negotiation of various living and non-living resource agreements, for example, were within this period. In comparison with other southern hemispheric nations, Chilean Antarctic policies have been infused with Pinochet's vision of Chilean geopolitical development (Jilberto 1991). In his annual address to the nation in September 1975, Pinochet recorded that: "The government has adopted measures and traced the lines of a policy designed to preserve our solid rights in the National Antarctic territory, now incorporated into the Antarctic province of the 12th region" (cited in Pittman 1981:1403). Government strategy in the 1970s and 1980s was committed to the geographical, resource and political integration of the CAT. As with his earliest geopolitical writings, a form of naturalized geopolitical discourse played a distinct role in conceptualizing Chile's regional development strategies (Pittman 1981). Chilean polar territories were consistently represented as an open space awaiting future territorial colonization and resource exploitation (Riesco 1985).

Under Pinochet's dictatorship, individuals attached to institutions such as the Instituto Naçional Antarctico de Chile and the Instituto de Geopolitico de Chile have contributed to a widely held political consensus that Chile had to protect its southern polar claims whilst projecting itself as a major southern hemispheric maritime power. Along with other southern hemispheric pariah states such as Argentina and South Africa, Pinochet's Chile was profoundly influenced by the politics of the Cold War and policies of anti-communism. Chilean activities in the Antarctic were not immune from these wider currents in spite of the provisions of the Antarctic Treaty and/or the protestations to the contrary from Chilean scientists or civilians. There was concern, for example, that Cuban and Soviet fishing boats in the South Pacific were engaged in spying throughout the 1970s and 1980s (Child 1988:198). However, conflict with Argentina over the Beagle Channel and tension over the 1982 Falklands/Malvinas War threatened, at one stage, to shatter strategic and political relations with Argentina.

The geopolitical basis of Pinochet's regional and developmental policies has been examined by a number of Chilean and American writers (eg Pittman 1981, Child 1985, Riesco 1985, Toledo and Zapater 1991). In line with other South American presidential figures such as Perón, Pinochet placed considerable priority on regional development and resource frontiers (Pittman 1988a). Proposals for regional development in the Antarctic have followed a well-worn path of previous episodes of territorial and resource colonization in places such as the Atacama desert and the southern parts of the Republic. The 1979 proposal for the "Regionalization of Chile" emphasized the long-standing geopolitical belief that Chile was a tri-continental nation composed of insular, polar and continental territories. Moreover, special developmental priority was to be placed on the geographical extreme regions of Chile: the Atacama (Region I) and the CAT (Region 12). Under this proposal, the Pinochet government was committed to two policy themes: the construction of new land and sea highways to the southern parts of the Republic and the promotion of the

Chilean Antarctic programme. This was widely perceived to be in keeping with his earlier arguments about the geographical continuity of the Chilean nation. President Pinochet's visit to the Antarctic in 1977 epitomized this widely held belief. As Luis Mericq has noted, "There at Marguerite Bay . . . he deposited an urn with earth brought from all regions of Chile, conveying that the CAT is united to continental Chile and only physically separated from it by the Drake's Passage" (Mericq 1988:102). In combination with changes to school education and military instruction, Pinochet's high profile speeches and polar visits have contributed to the campaign to raise the Antarctic consciousness of the Chilean people.

Chilean Antarctic policies under President Pinochet were directed in three distinct directions: first, as already noted, Chile's negotiations over resource management were linked to a robust defence of the Antarctic Treaty System. Moreover, resource exploitation coupled with effective environmental management had been a dominant theme within the Chilean ATCP delegation. Chile's initial support for the CRAMRA proposals was indicative of this policy concern. The second direction was the development of the Chilean presence in the CAT. The creation of the airfield in the Teniente Marsh greatly facilitated military and scientific access to the base. Chile's claim to being the "gateway to the Antarctic" had been underlined by promoting polar tourism and supporting major international projects such as the ozone depletion study carried out by the National Oceanic and Atmospheric Administration. Third, Chile had been involved in a series of tense geopolitical relations with Argentina over access to the Beagle Channel, control of the Drake's Passage and resource exploitation in the Southern Ocean. As mentioned earlier in the Argentine chapter, the 1982 Falklands/Malvinas War also raised some difficult issues regarding Anglo-Chilean co-operation during that conflict. From the Chilean perspective, an Argentine victory in the South Atlantic might have heightened the possibility of war being declared over the Beagle Channel dispute. Ironically, in 1978 the Chileans had agreed the Argentine–Chilean Act of Puerto Montt which confirmed the legal and political importance of a South American Antarctic (Beck 1990b:103).

Geopolitical reasoning was paramount during the Beagle Channel dispute as this particular geographical feature had been invested with strategic and political value in the latter part of the present century. From an Argentine perspective, the dispute resided in a failure of the Chilean state to recognize Argentina as a South Atlantic power with a distinct geographical and political jurisdiction. Under the so-called "bi-oceanic principle", Chile was assigned to the South East Pacific by Argentine geopolitical writers (Child 1988, Pittman 1988b). Regions such as the Drake's Passage and the Beagle Channel have been encoded as strategically important not only in terms of their possible use for the international community (eg an alternative shipping route to the Panama Canal) but also because the demarcation of the South Atlantic and the South Pacific was crucial to assigning geographical responsibilities.

The 1984 Treaty of Peace and Co-operation was, therefore, a milestone for Argentine–Chilean relations as it resolved the conflict (Morris 1984). Whilst the threat of war over access and ownership of the Beagle Channel was averted, there remained in the mid-1980s fears that conflict might yet occur over the south-western portion of the Atlantic, the Antarctic peninsula, the Drake's Passage and/or the south-eastern portion of the Pacific Ocean. Speaking at the United Nations General Assembly in 1984, Chilean Ambassador Fernando Zegers reiterated the familiar Chilean position over the Antarctic and related seas and islands:

> Chile's ties with Antarctica have to do with our geographical position, our history and mission. We are less than 500 miles from the Antarctic continent. We are the country nearest to it; and our country reaches as far as the South Pole. Chilean planes reach Antarctica in two and a half hours from Punta Arenas. Chile maintains three permanent bases there throughout the year, an intense scientific work, an effective occupation with the meteorological network and with tourism . . . Chile is therefore, by definition an Antarctic country; it is that because of historic missions, its geographical position, our presence there and the activities of our men. (cited in Naraynana and Priya 1986:185)

Boundaries remained a sensitive issue throughout the 1980s as Argentine and Chilean diplomats attempted to negotiate new lines over the sea and ocean floor. At the same time, Chilean geopolitical writers and political leaders developed elaborate visions of maritime and polar space.

During the 1980s, maritime consciousness continued to occupy a prime position within official Chilean discourses on the CAT and Southern Ocean because of events related to the 1982 Law of the Sea Conference and the resource issues dominating the business of the Antarctic Treaty System. The Convention on the Law of the Sea incorporated the concept of the "common heritage of mankind" into its discussions on the ocean floor and the high seas. The Chilean delegations were undoubtedly hostile to the common heritage comparision of the Antarctic with the ocean floors. Pinochet's geopolitical vision of Chile and its extra-continental responsibilities could not have tolerated such a position and this was reflected not only in Chilean statements to the ATCP meetings and UNGA debates but also through its scientific programmes and related events such as the hosting of a major international meeting in 1982 at Chile's Rodolfo Marsh base.

Chilean Antarctic Territory in the aftermath of Pinochet

In 1989, democratic elections were held in Chile which resulted in the formal removal of President Pinochet and the election of President Patricio Aylwin (Marcella 1990). The immediate effect on Chilean Antarctic policies appeared marginal given the continued reliance on the armed forces for polar logistical support and the frequent statements concerning Chilean sovereignty over the CAT. The long-standing association of figures such as Oscar Pinochet de la Barra and Fernando Zegers permeated Chilean polar policy making and the ATCP meetings (Pinochet de la Barra 1995). Another reason was the overwhelming consensus that existed amongst policy makers, the media and academics that the CAT remained an essential territorial component of the Chilean Republic (Child 1990). Finally, it has been argued that continued membership of the ATS remains the best option for ensuring Chilean sovereign rights to the region and the effective management of the polar environment (Pinochet de la Barra 1995). President Aylwin has committed Chile to: "maintaining there an active presence . . . the protection of its ecology and the maintenance of its demilitarized nature . . . the preservation of the current regime of the Antarctic treaty . . ." (Aylwin 1990:12).

In the post-Pinochet era, the management of polar tourism and the protection of the Antarctic environment have become the most important issues within Chilean Antarctic policies. For the first time in 40 years, these issues have received explicit priority over the traditional themes of national sovereignty, resource exploitation and territorial control. It was argued that until recently: "There was little space within such a political horizon for questioning the benefits of industrialism and development, since both extremes of the political spectrum were struggling around the distribution of resources. The question of the biospheric impacts of industrialization, urbanization, modernity and progress was simply not on people's minds" (Rojas 1994:96–7). The emphasis in that shift has been illustrated by the expansion of Chilean environmental movements such as the Comité de Defensa de la Flora y Fauna de Chile (CODEFF), the only Latin American NGOs directly concerned with the Antarctic. Founded in 1963 but not actively involved in polar matters until 1990, CODEFF, according to one of its key members, was organized on the basis of: "operat[ing] on several levels; we lobby a lot, work through communications with the public and legal actions against polluters, companies and government . . . We use public action as a tool for influence, and we have a lot of international level contact in order to get results" (Verscheure 1992 cited in Darby 1994:65). The long-term significance of CODEFF remains to be assessed but environmental issues did feature as a concern during the 1993 presidential elections, and it is notable that CODEFF joined the official Chilean delegation to ATCP meetings in the 1990s (Darby 1994:65). This is in marked contrast to the Argentine and other South American delegations which have shown little interest in the concerns of environmental movements.

The development and management of tourism is a "front burner issue" (Beck 1994a:375). The growing number of tourists visiting specific sites on the Antarctic peninsula is placing ecological pressure on these environments. It has been estimated that around 8000 tourists visit the Antarctic every year (Enzenbacher 1992). Chile was prominent in opening up sea-based tourism via Punta Arena (Hall and Johnston 1995). The control and/or management of tourism present some vexing legal and political questions such as the regulation of commercial operators and legal liability. This has been compounded by the fact that the ATS has not yet carried out either a systematic investigation of "tourist impact" or established a set of regulations governing polar tourism (Chaturvedi 1996).

During the 1991 ATCP meeting in Bonn, the Chilean delegation submitted a working paper proposing that a separate annexe on tourism be added to the Protocol on Environmental Protection. The Chilean proposal (8 October 1991) included provisions for the notification of tourist visits, the developement of rules of conduct for visiting tourists, the requirements to be fulfilled by tour operators, the protection of historic sites and the development of an insurance/liability regime. Whilst this was opposed by some states such as Australia, on the grounds that the Protocol already encompassed all human activity, it did reflect Chilean concerns that unregulated tourism might endanger the polar environment. Lorraine Elliot has argued that a separate annexe on tourism would be resisted by NGOs because it could be used by ATCPs to veto their activities in the Antarctic (Elliot 1994b:202). Whilst the 1991 meeting was unable to secure a consensus over the regulation of tourism, the ATCPs did arrange a subsequent informal meeting in Venice in November 1992 in order to negotiate this issue. Chile, in conjunction with other states such as Germany and France, resumed its arguments for a separate annexe on tourism. The parties failed to secure agreement on this issue given the opposition of Australia, the UK and the USA to such a suggestion.

The 1994 ATCP meeting in Kyoto eventually constructed a general agreement on the regulation of tourism. Under Recommendation 8, the ATCPs agreed to the following obligations for tour operators and organizers: prior notification and reports of activities to relevant authorities and/or parties; environmental impact assessments of planned activities on the continent or sub-islands; prevention of waste disposal of banned substances; respect for scientific research and support for existing measures for conservation and management; and an effective response to any environmental emergencies such as oil pollution (Chaturvedi 1996:222). However, it is apparent that the ATCPs have not reached agreement over whether the Protocol should be altered to include a separate annexe or whether there is even sufficient data on tourist activity to construct satisfactory measures for the regulation of polar tourism. Indeed, the Chilean delegation submitted an agenda item for the Kyoto meeting complaining that the slow resolution of this issue was "of particular concern for countries which surround the continent of the South Pole and whose ecosystems are dependent and closely associated with the Antarctic ecosystem" (ATCP 1994). Chile later proposed the creation of a Coordinating and Implementing Group to examine the case for a separate annexe to the Protocol. This was not accepted by the other ATCP parties in their April meeting, but the Transitional Working Environmental Group will continue to be instrumental in implementing the provisions of the Protocol.

The last five years have witnessed a change in Chilean attitudes towards the Antarctic in the aftermath of the Protocol. Whilst the tensions of the 1980s regarding the mineral issue and the challenge of the Developing World have mellowed, the Protocol has not permanently resolved the territorial and resource status of the continent. Protecting their rights to the CAT is likely to remain a policy priority, especially as some parties such as Australia have argued for the establishment of a permanent secretariat to handle the administration of the ATCPs. Chilean opposition to a secretariat is rooted in long-standing concerns over further internationalization of the Antarctic and the possible diminution of Chilean national sovereignty. This should not, however, imply that Chile has been reactionary in the face of developing measures for the environmental protection of the Antarctic or in devising measures for the regulation of tourism. An interesting development in the last two or three years has been the promotion by the Chilean delegation to the ATS of bipolar links with the member states of the Arctic Environmental Protection Strategy such as Canada, Finland and Sweden (Sugden 1982, 1989). During the 1994 Kyoto ATCM, for example, Oscar Pinochet de la Barra submitted a Chilean paper which called for the possibility of "a future relationship between the systems for the protection in the Arctic and Antarctic" (ATCP 1994). The 1994 ATCM later acknowledged the importance of relating regional conventions affecting the Antarctic to the experiences of the Arctic. The protection of fragile ecosystems, the implementation of environmental impact assessment and the monitoring of pollution were declared common problems to both regions by the ATS in 1994 and again at the 1995 ATCM in Seoul.

The Chilean proposal for a formal Arctic/Antarctic interchange has been a decisive recognition that the environmental protection of the Antarctic is linked to regional and global environmental management. The Chilean chairmanship of the Valdivia Group during 1995–6 was dominated by the challenges posed by dealing with regional problems such as ozone depletion and the inherent difficulties in formulating common environmental strategies. Accordingly, the Chilean armed forces also recognize that Chilean environmental security poses different challenges to more traditional concerns of territorial boundaries and resource exploitation. In an interview with a South African analyst, Chilean Naval Under-Secretary Pablo Cabrera was reported in October 1996 as noting that:

> We feel that our borders are no longer land borders; in this globalised age we are bordered by oceans. Environmental problems cross these borders with impunity. Through the oceans we feel we are close to many countries and can face common problems together with other nations with similiar policies . . . As middle-sized countries, we play key regional roles. We have to work together in trade and in the sharing of technology. There are many things we can do together through the fragile environment which is the oceans. (cited in Mills 1996:11)

The promotion of a common course of action and co-operation in the Antarctic and in related polar science issues has superseded the maritime-focused geopolitical thinking of earlier periods.

Summary

The Chilean claim to the Antarctic, based on the historical and geographical evolution of the Republic has been dominated by concern over territory and resource control. In keeping with South American history, the uncertainties of the post-independence era, following the collapse of Spanish authority transformed the Antarctic into a contested space within the Chilean geopolitical imagination. The overlapping claims with Argentina and Britain have been an enduring source of conflict and ensured that the traditional geopolitical agenda favoured by military writers has predominated in the last 40 years. In the aftermath of President Pinochet, however, recent Chilean polar policy has been more supportive of international scientific co-operation and the development of environmental measures for the protection of the Antarctic environment. Discourses on environmental security have become more significant within Chilean political and media circles as Chile has participated in regional organizations such as the Valdivia Group and the Reunion de Aministradores de Programs Naçionales Antarticos Latinamericanos in the 1990s.

As a founder-member of the ATS and a claimant state Chile has participated in the development of measures and regulations for resource conservation, environmental protection and the regulation of tourism. In comparison with other southern hemispheric nations such as Australia and Argentina, Chile has also been a strong advocate of her territorial and resource rights in the polar region. The Instituto Antartico Chileno under the leadership of Pinochet de la Barra was a prominent advocate of Chilean Antarctic sovereignty. Likewise, the scientific research programmes supported by the armed forces performed a well-understood role of "effective occupation". As the first mayor of a Chilean Antarctic community, Daniel Contreras concluded, "Our presence here is the best guarantee of Chilean rights in this territory and of its permanent interest in keeping our sovereignty" (*Boletin Antarctico Chileno* 1984:61). However, in alliance with other southern hemispheric nations such as New Zealand, it is likely that Chile will play a positive role within the ATS and seek to implement the provisions of the Protocol. Public discourses on the Antarctic and Southern Ocean are beginning to change away from the narrow confines of territorial sovereignty towards environmental security and ecosystem management (Pinochet de la Barra 1995).

6

India and the Antarctic

The first Indian polar expedition, led by Dr S.Z. Qasim, left Goa in December 1981 and landed at Dronning Maud Land in January 1982. As Peter Beck has noted, "Logistical problems reduced the length of stay from the planned 25 to 10 days, but a wide range of scientific observations was carried out. The expedition returned in February 1982, when Mrs Indira Gandhi told the *Lok Sabha* (Lower House in Delhi) that the main objective had been to study the meteorology and other conditions of Antarctica, which are believed to control the monsoons and to influence the climate of the Indian Ocean region" (1983:106). Shortly afterwards, despite many years of public opposition to the Antarctic Treaty System, the Indian government formally accepted the provisions of the 1959 Antarctic Treaty and applied for Consultative Party status. This was later accepted by the consultative parties on the basis of India's planned scientific research rather than existing polar achievements (cf. West Germany). According to the former Head of the UN division at the Indian Ministry of External Affairs, T.P. Sreenivasan, India's accession to the Antarctic Treaty System in 1983 reflected a number of interrelated concerns of the Indian government: to make the ATS more receptive to the views of Third World states, to distribute the benefits of scientific research in the Antarctic, to ensure that discussions in the UN would not necessarily lead to the dissolution of the ATS, to protect India's interests in any future discussions over a mineral resources regime and, finally, to promote the protection of the Antarctic environment (Sreenivasan 1994; also Herr and Davis 1994).

Since the creation of the Indian state in 1947, the Antarctic and the Southern Ocean have occupied the minds of successive political leaders. Political interest in the Antarctic has been intimately linked with India's geostrategic, resource and meteorological interests in the Indian Ocean

(Haggman 1994). Whilst regional trading and security interests in South Asia may have been more important in the post-independence era, Indian foreign policy has consistently devoted considerable time and resources to oceanographic research, fishing and naval strategy. According to the Indian geopolitical writer Sanjay Chaturvedi:

> The Indian view [of the Indian Ocean and Antarctic region] is framed in the east by the fragmented Indonesian archipelago, scantily developed and sparsely populated West Australian coast and in the west by the multi-state coastline of the South Arabian Peninsula swooping down south along the East African coastline. To no other virtually contiguous frame of coastline in the world is Antarctica so exposed. Since oceans today link (rather than obstruct), India is formidably linked to Antarctica by nature; for a technologically growing India the geographic reality and value of Antarctica grows the more, steadily and surely. (1990:161–2)

Furthermore, Indian officials have paid considerable attention to the organization of the international politics of the Antarctic even though India's accession to the Antarctic Treaty System in 1983 reversed earlier reservations about that political system.

This chapter relates the earliest Indian pronouncements on the future of the Antarctic and proceeds to analyse her political, resource and strategic interests in the Indian Ocean. Thereafter, the more recent activities of Indian Antarctic research programmes and the role of the Department of Ocean Development are investigated. It will be stressed that India's involvement in the Antarctic was shaped by a variety of factors such as resource concerns, strategic interests, national prestige and regional hegemony. In the process, the subsequent accession to the ATS coincided with the evolving debates on the "Question of Antarctica", CRAMRA and UNCLOS III. India's future role within the ATS is further complicated by the prospect of Pakistani participation in this political system.

Post-independence India and control of Antarctica

The commitment and vision of Jawaharlal Nehru to build a modern and developed nation has been a central feature of Indian domestic politics. As Stuart Corbridge notes, "Nehru was able to personify the developmental ambitions of a unitary India . . . empowered [through control of the Congress Party] to push for this development by means of a technocratic state apparatus and a Planning Commission which would consist of the country's top economists and politicians" (Corbridge 1993:184). For most commentators, India as the major regional power in South Asia proved to be one of the most successful of the so-called "developing nations". In 1947, India was regarded as the 10th largest industrial power even though it did not possess a manufacturing industry at the time of independence (Corbridge

1993:181). The post-colonial state became committed to a political economy of development which would encourage state-led policies of industrialization, usually identified as forms of import-substitution industrialization (ISI). With the financial and technological aid of both superpowers, India's industrial plans were based on capital goods substitution and the development of a local industrial infrastructure.

In a geopolitical setting, India was rapidly encoded as a strategically important state in the Cold War struggle for hegemony in South Asia. The United States made substantial transfers of food aid to India in the 1950s, whilst the Soviet Union provided military and financial aid in the form of energy projects and weapons procurement (Hewitt 1992:12). Against this backdrop, Nehru was able to play an active role in Third World and non-aligned politics. In conjunction with other high profile Third and Second World leaders such as Nasser of Egypt and Tito of Yugoslavia, Nehru contributed to a new politics of non-alignment (Willetts 1978).

At the 1955 Bandung Conference, recently decolonized African and Asian nations met for the first time to discuss their collective interests as both ex-colonies and under-developed economies. The main preoccupation of the group was the creation of a movement of non-aligned nations capable of influencing international affairs. Born out of a collective struggle against the legacy of colonialism, the development of a non-aligned ideology was forged in the midst of Cold War alliances. As Nehru noted in 1957; "The Cold War is based not only on hatred and violence, but also upon a continuous denunciation, on picking out the faults of others. I tried to reverse this process, even when I differed radically from those that I addressed" (cited in Hewitt 1992:16).

These ideologies of non-alignment were employed to serve the regional powers (especially India and Egypt) in their bid to cultivate good relations with both superpowers. The ongoing discussions over the future of the Antarctic amongst the nations of the advanced world at the expense of the emerging Third World seemed to epitomize the accusations of the Non-Aligned Movement (NAM). India's moves towards the Antarctic were indicative of Nehru's desire to expose the political and intellectual mind-sets of the leaders of the First World to decolonization. Whilst India was determined that it should have an impact on mainstream international politics, other general principles of non-alignment such as mutual non-aggression and mutual non-interference have not always been implemented by Indian governments anxious to secure their territorial borders with China and Pakistan (Hewitt 1992).

In October 1956, a formal proposal entitled the "Peaceful utilisation of Antarctica" was submitted for inclusion in the 11th agenda of the United Nations. A letter from A.S. Lall, India's permanent representative, to the secretary-general of the United Nations clearly highlighted the perceived future importance of Antarctica for the general welfare of the world. Lall pointed out that: "any disturbance of the equilibrium of natural forces in this area might lead to incalculable consequences . . . involving the deterioration of the conditions for human and other forms of animal and plant life" (cited in Dey 1992:174). The Indian government accordingly suggested that the United Nations was the appropriate venue for discussions on the future of the Antarctic continent given three major issues: the resource value of the continent, the climatic significance of Antarctica and the environmental significance to the globe.

> Antarctica, a region covering about six million square miles of territory, has considerable strategic, climatic and geophysical significance for the world as a whole. With the development of rapid communications, the area might shortly come to have further practical significance to the welfare and progress of nations. The mineral wealth of the landmass is believed to be considerable and its coastal waters contain important food resources . . . the Government of India considers that in order to strengthen universal peace it would be appropriate and timely for all nations to agree and to affirm that the area will be utilised entirely for peaceful purposes and for the general welfare. (UNGA 1956)

This proposal met with strong opposition from India's commonwealth partners (who had not been consulted) and other claimant states. A Chilean press communiqué released on 23 February 1956 noted that: "The Chilean government considers inadmissible that this matter, relating to territories over which Chile has sovereign rights which do not admit discussion, be brought to the attention of the world organisation for debate." The Chilean foreign minister, Enrique Gajardo, wrote in the Chilean newspaper *El Mecurio* (22 February 1956) that the Indian proposal was also contrary to the charter of the United Nations which prohibited outside interference in internal affairs. The Canadian records of the Department of External Affairs record the hostility of the British, Australian and South African governments to Lall's proposals. The Australian government was reportedly the most vociferous critic of India's plans to "internationalise" the Antarctic. The Canadian High Commissioner in Canberra reported the words of the Australian foreign minister after a meeting in February 1956: "I could not see the relevancy of making the continent an international trust territory since there is no indigenous population whose interests need to be safeguarded" (National Archives of Canada, Department of External Affairs 1956). The reaction of the New Zealand government was less "heated" because of existing sympathies for a new international political settlement for the region. A telegram from the Canadian High Commissioner in New Dehli to the Department of External Affairs in Ottawa recorded, "We too find the Indian initiative surprising. Our initial reaction is to see two possible motives: first, a genuine scientific interest . . . Second, one with political implications along the lines of a public proposal made in January in New Zealand by the leader of the opposition, Mr Nash, that Antarctica be declared an international trust territory under the complete control of the United Nations Assembly" (National Archives of Canada, Department of External Affairs 1956). After a series of meetings of Commonwealth leaders in London and New York, it was concluded that the Indian government should be dissuaded from such an initiative.

The Indian proposal for the inclusion of an item on Antarctica on the agenda of the 11th session of the General Assembly had not, however, materialized in a political vacuum. In the period between 1948 and 1956, various proposals had been presented for the future management of the Antarctic: the American government had suggested a trusteeship arrangement amongst eight interested parties in 1948; the Chileans suggested in 1950 a "standstill agreement" which simply protected the sovereign rights of claimant states; the New Zealand government had proposed a UN-supervised condominium which would only regulate scientific research and living resource management and, finally, the Australian government proposed an advisory commission which would merely guide interested parties in possible collective management. None of these existing proposals undermined existing claims to territorial and maritime rights nor sought to involve a wider managerial segment of the international community. The Indian proposal clearly invited suspicion from Commonwealth and South American nations precisely because it proposed that this uninhabited territory should be regulated by the United Nations rather than a select group of states already involved in the polar region (Dodds 1996c).

Commonwealth opposition forced the Indian government to withdraw this request on 14 November 1956. Nevertheless, the Indian initiative had caused political "shock waves" to be sent through the foreign ministries of the old Commonwealth. It was widely believed that India's concern for the Antarctic might be linked to a number of connected factors:

(1) fears over possible nuclear testing and general militarization in the polar region;
(2) exclusion from the current pattern of resource exploitation in the Southern Ocean;
(3) a desire to raise India's international profile in the United Nations and other international forums such as the NAM.

Whatever the motivations, the old Commonwealth nations, including Canada, were deeply concerned that this Indian initiative would either disrupt the delicate polar negotiations with the United States and/or encourage the Soviet Union to declare an Antarctic territorial claim. In February 1958, India as a non-Antarctic participant in the IGY, again raised the possibility of placing the political status of Antarctica on the agenda of the United Nations General Assembly. In the formal memorandum it was stressed that this discussion should be accomplished: "without any nation renouncing such rights as it may claim in Antarctica, or claims to sovereignty, or other rights consistent with the Charter" (cited in Bush

1988:510). The Indian premier Jawaharlal Nehru later acknowledged in the Indian Parliament in May 1958 that:

> We are not challenging anybody's rights there [in Antarctica]. But it has become important, more specially because of the possible experimentation of atomic weapons and the like, that the matter should be considered by the UN . . . the fact that Antarctica contains many very important minerals – especially atomic energy minerals – is one reason why this area is attractive to various countries. We thought it would be desirable to have a discussion about this in the UN. (cited in Dey 1992:173)

The softening of tone from the 1956 memorandum was striking in so far as it acknowledged that claimant states were unlikely to relinquish these claims. Nehru's speech also hinted at the wider strategic and resource interests that were of official concern to New Delhi but India was later persuaded to withdraw its request by both the Commonwealth governments and the United States who were already involved in preparatory meetings for the Washington Conference (Quigg 1983:166).

This second Indian proposal for an item on Antarctica to be included on the agenda of the 13th session of the General Assembly reflected the changing circumstances of the international politics of the Antarctic. By January 1958, most interested parties had concluded that a new political arrangement would have to be negotiated after the ending of the IGY. Newspapers such as the *Sunday Observer* (26 January 1958) carried the reported comments of Mikhail Somov, a major figure in Soviet polar exploration, that: "Russia wanted to be represented at any conference which decided the future sovereignty in the Antarctic." By the middle part of 1958, the Commonwealth states and the United States had conceded that the Soviet Union was not likely to leave the Antarctic at the end of the IGY. India, in conjunction with other SORS, was concerned lest any subsequent agreement over the future of the Antarctic failed to recognise that these states had specific concerns such as the impact of nuclear testing on the polar continent and its effects on neighbouring continents.

The invitations to a conference on the future of the Antarctic was organized by the United States and extended to the participants of Antarctic research during the IGY. Whilst India had participated in the scientific programmes of the IGY, it had not been involved in polar research. Nehru's vision for a modern India meant that this technological and scientific involvement had nonetheless been substantial and it had encompassed the Indian continent as well as the Indian Ocean (Chaturvedi 1990, Corbridge 1993). According to official Commonwealth records, the Indian government subsequently raised no objections to their exclusion from the Washington Conference. The New Zealand government of Walter Nash played a key role in reassuring Nehru of the benign political intentions of the Antarctic Treaty parties. In his letter to Walter Nash, Nehru confirmed that his interest in the Antarctic was based on the following:

> I have been deeply concerned by the increased international tension following the events of the Middle East. In this atmosphere I am reluctant to do anything which might create the slightest friction. In view of what you have said I am asking our delegation not to press for a discussion of Antarctic [*sic*] at the forthcoming session of the General Assembly. (National Archives of New Zealand, Department of External Affairs 1957)

The Parliamentary Under-Secretary to the Indian Minister of External Affairs, Sadath Ali Khan, later confirmed to the Lok Sabha in September 1958 that India would neither raise the issue of Antarctica again at the UN nor seek to participate in the series of preliminary meetings to be held at Washington.

The Indian proposals to discuss the future of the Antarctic in the United Nations had provoked considerable debate over membership of the Antarctic Treaty System and the relationship this regime would have with the United Nations. The Soviet Union, supportive of India's proposals in 1956–7, would not have objected to Indian participation at the Washington Conference. At one stage in January 1958, even the Macmillan government considered admitting India as a second-tier member of any new international regime negotiated in the USA. A potent reason for such a policy option was the Indian representative at the UN, Krishna Menon, who had been highly active in raising attention to the Antarctic issue and thereby forcing Commonwealth representatives, fearful of a widening of discussions over the future of Antarctica, to seek a solution. Later, in comments attributed to Arthur Lall, subsequent provisions of the Antarctic Treaty, especially the banning of nuclear activities, were considered to have covered the major points of possible objection. After the Treaty's ratification in 1961, the persistent presence of the South African government in the ATS was considered objectionable on the basis of a general rejection of apartheid (Chaturvedi 1990).

Indian interests in the Indian Ocean: resource, strategic and geopolitical interests

Indian interest in the Antarctic lessened in the 1960s, as sub-continental territorial conflict and developmental issues assumed a greater importance. In the case of the former, the Indian state found itself involved in wars with the Chinese in 1962 and with Pakistan in 1965 and 1971 over territorial boundaries and zones of influence. The Indian government also became increasingly preoccupied with the politics of development, epitomized by the embracing of the Green Revolution. American food aid to India had ended in 1964, as the USA sought to renegotiate trading relations due to problems over India's domestic food supply and complicated by growing Soviet influence in Indian foreign policy. American aid was now subject to the following conditions: no Indian support for Soviet-backed North Vietnam, an end to the policy of non-alignment, devaluation of the rupee and an endorsement of foreign investment into specific regions and industries (Corbridge 1993:188).

Geopolitical considerations were conclusive factors in shaping India's foreign relations. Whilst Indian scientists participated in the 1960–1 Australian expedition and the 1971–3 Soviet Antarctic expedition, the international politics of South Asia and the machinations of the Cold War continued to dominate Indian foreign policy at the expense of further involvement in the polar region (Hewitt 1992:66–8). The Indian Ocean, however, was considered a strategic priority for successive Indian governments in the post-independence era. Growing political and legal interest in the sea after 1950, coupled with Western dependence on the Suez Canal and the Persian Gulf oil routes, meant that the Indian Ocean was transformed from a strategic backwater to a place of major international concern (Puri 1986, Chaturvedi 1990). Political arguments over the importance of sea lanes, territorial waters and resource distribution were shaped by the ongoing negotiations associated with the United Nations Convention on the Law of the Sea (UNCLOS). The 1958 Convention on the Law of the Sea was the first major post-war deliberation on the legal and political status of territorial seas, continental shelves and the high seas. Although, as Martin Glassner noted, "the breadth of the territorial sea and the establishment of exclusive fishing zones" was left undecided, the first conference had succeeded in setting up some basic principles of international maritime law (1993:474). The second UN conference in 1960 was less successful and failed to secure agreement over the geographical breadth of the territorial sea. Thereafter, the international community struggled to reach agreement over the legal and political status of the deep seabed and the ocean floor. The deliberations of UNCLOS III were largely taken up with concerns first raised in 1967 by the UN *ad hoc* committee on the Peaceful Uses of the Seabed and Ocean Floor Beyond the Limits of National Jurisdiction, better known as the Seabed Committee.

The sustained interest of the Indian state in the resources of the Indian Ocean was interpreted by some scholars as a product of the politically fraught 1970s. Domestically, the Indian government of Indira Gandhi was struggling to hold together Nehru's federal and democratic vision of India. The Indian economy became increasingly weaker as the government was forced to print ever more money in order to fund social programmes and finance industrial development. The economic pressures also fed regional and political tensions as new challenges to Nehru's dream of a secular India emerged in places such as the Punjab and Assam (see Akbar 1985). On the international scene, growing concerns over superpower machinations in the world oceans, coupled with neo-Malthusian fears of environmental catastrophe, were placing new pressures on resource exploitation and distribution (Hurrell 1994). As S.S. Bhattacharya concluded:

> In the Indian Ocean, the US after developing Diego Garcia as a fully-fledged modern nuclear base is now strengthening its position in the north-western sector of the ocean by creating a rapid deployment force. In the Pacific, it has taken adequate steps to strengthen its naval positions in the western sector of the ocean . . . as a counterbalance to the growing Soviet influence in the region . . . The Soviet Union has economic interests in the Afro-Asian region. In maintaining and developing its political and economic interests with this region, the sealanes of Southeast Asia play a significant role. (Cited in Chaturvedi 1990:164–5)

As early as 1964 the Indian government had pressed for the creation of the Indian Ocean as a zone of peace, fearing superpower competition and possible nuclear confrontation (Government of India, 1987).

The interrelated pressures of resource distribution, decolonization, superpower rivalry, mineral exploitation and the endeavours of scientists (such as the International Indian Ocean Initiative 1959–65) to discover more about the ecosystems of the oceans became ever more time-consuming in the 1970s (Joyner 1990a, Hurrell 1995). India and South Africa were increasingly concerned that access to oceanic resources should be available to all nations rather than simply the preserve of the technologically advanced world. The mineral potential of the Antarctic and the Southern Ocean became a favoured topic of debate for the Indian strategic community, whilst the lengthy proceedings of UNCLOS III (1973–82) formed the backdrop to many of these discussions, dominated as they were by uncertainties over the ownership of common resources and the demands of the Group of 77 that these deliberations be set within the context of a New International Economic Order (Bandyopadhyaya 1979). The decision of the NAM in 1976 to circulate a draft resolution on the "Question of Antarctica" was considered a diplomatic triumph by Indian geopolitical observers coming as it did two years after the UNCLOS meetings in Caracas in 1974 (Glassner 1993:476).

The NAM initiative, which had been proposed by Sri Lanka, suggested that the exploration and management of Antarctica's resources should be carried out in a fashion that would secure maximum benefit for the global community rather than a select group of developed nations. Whilst the resolution did not appear in the final proceedings of the Colombo NAM conference, the Indian government ensured that this issue was brought to the attention of the General Assembly in 1976 (Mohan 1984:164). The rationale for this form of internationalism was consistent with wider agendas established by Indian foreign policy based on non-alignment, global economic reform and expanding the political influence of the Third World. The following year, Indira Gandhi ordered a secret study on Antarctica with special reference to India's national interests (Nawaz 1986:73). The outcome of this research was the subsequent planning and execution of India's first polar expedition in 1981 (Dey 1992, 1994).

In spite of growing official and public interest in the Antarctic in anticipation of this major polar expedition, pressures associated with the political economy of development, the future of the democratic polity and the territorial integrity of the Indian state dominated the political agenda during the Cold War. India's international relations were fundamentally shaped by periodic co-operation with the superpowers and an uneasy coexistence with its neighbours in South Asia (Hewitt 1992). In particular, Indo-Pakistani rivalry continued unabated as a dangerous element in South Asian politics. Uncharitable observers may even interpret Indian involvement in the Antarctic in the 1980s as a product of this competition, given existing tensions in the fields of nuclear weapons programmes, territorial boundaries and ethnic irredentism. Notwithstanding the regional complexities of South Asian politics, Indian interests in the Southern Ocean have also been shaped by the activities of the superpowers:

> In spite of their best efforts, they have not been able to stop the big maritime powers from pursuing their rivalries and playing their war games in the vicinity. Instead of developing into a zone of peace, the Indian Ocean has become a region of conflict. The only way for the small littoral and hinterland states of the Indian Ocean to protect themselves is through regional co-operation. ASEAN, SAARC and the GCC are steps in the right direction. But we still have a long way to go. (Anand 1988:44)

This message was partially reiterated by Indira Gandhi in her announcement to the first Indian polar expedition team in 1982: "The Indian Ocean links India to Antarctica. The entire area is of deep interest to us, and ocean studies are of vital importance" (cited in Dey 1992:177). Geographical and oceanographic continuity apparently was used to define natural interests and to compel India to undertake Antarctic endeavours, a line of geopolitical reasoning that has striking parallels with South American states such as Argentina and Chile.

India in the Antarctic

In December 1981, the first Indian polar expedition set sail from Goa with the expressed purpose of studying the environment and meteorology of the "Norwegian sector" of the Antarctic continent (see Map 6). To achieve this, an unmanned solar powered station called Dakshin Gangotri was established in January 1982. The symbolism of the moment was not lost on the Indian government – in English, the base's name would be the Southern Ganges. The iconographic significance of the name (with reference to the mythical Himalayan source of the Ganges see Larus 1984:46) should be acknowledged. Indira Gandhi extolled in February 1982 that the expedition offered: "One more proof, if such be needed, that Indian scientists and technologists have the capability to undertake the most hazardous and complex tasks . . . In undertaking this advanced work India has now joined a select band of countries" (cited in Dey 1992). The landing of the 21-man Indian expedition was the culmination of a long period of both strategic interest in the region as well as an indifference to the international political mechanism which had governed Antarctic politics (Beck 1986).

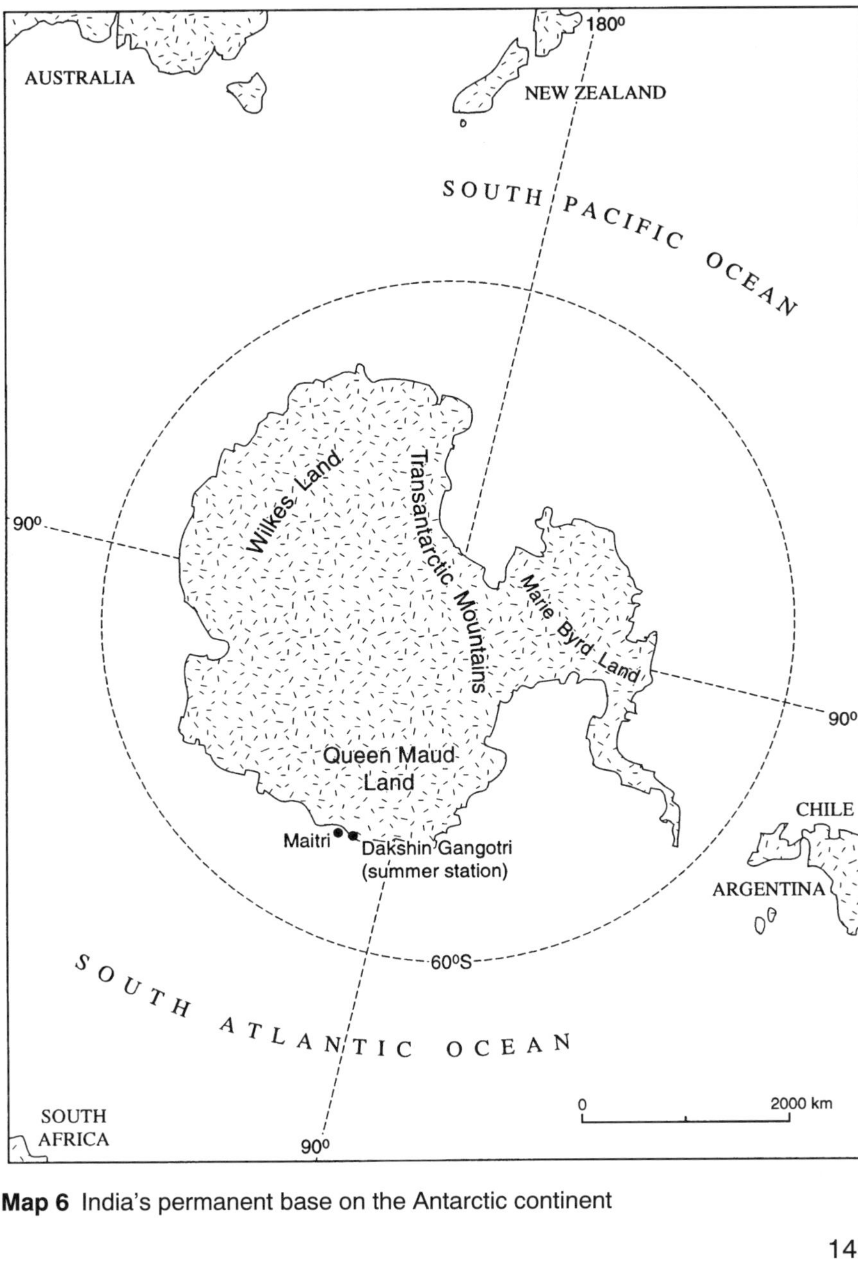

Map 6 India's permanent base on the Antarctic continent

This first Indian station was expected to produce a continuous record for one calendar year of the local weather patterns which would then be collected by the second expedition, Gangotri II. The Indian government had expressed a desire to test the suitability of Indian radio and weather-collecting equipment in the sub-zero temperatures of the Antarctic. These "equipment tests" would be invaluable as they were related to their ongoing territorial struggles with Pakistan in the mountain ranges of the Kashmir. Expedition members had been dispatched to Gulmarg in the Himalayas in order to participate in a cold weather acclimatization programme during October and November 1981 (Government of India 1982). Notwithstanding these observations, the expedition was undoubtedly funded because the Gandhi government was convinced that the timing was appropriate given the UNCLOS negotiations and the secret minerals debates within the ATS. Indira Gandhi's long-standing enthusiasm for an Antarctic project was further aided by international political and financial circumstances (Larus 1984:45–7). The Indian economy was considered stable, if not expanding, and the ongoing territorial disputes with Pakistan had lessened in their immediate political salience.

While the planning stages for Operation Gangotri were finalized in New Delhi and Goa, a major governmental re-organization had been approved by the Gandhi administration. The Department of Ocean Development (DOD), created in July 1981 under the president of India's Notification Number CD800/81, was given the responsibility of administering Antarctic expeditions (Dey-Nuttall 1994). Initially, it operated under the cabinet secretariat and then later in March 1982 became a separate department under the authority of the prime minister and the minister of state for science and technology, ocean development, atomic energy, space and electronics (Dey 1992:179). The first *Annual Report* states, under a section entitled "International Collaboration and Antarctic Programmes" that the DOD: "will be able to advise the Department on various matters relating to foreign collaboration, including the various types of programmes for providing assistance to India. It should also be able to handle the Antarctica project and its various implications (scientific, economic, political and legal) effectively and advise the Department accordingly" (Government of India 1982). The Indian armed forces participating in government decision making were instrumental in establishing the logistical support centre at Goa. To complete preparations, an independent Antarctic Study Group, chaired by the director-general of the Council of Scientific and Industrial Research, was established to advise and evaluate the scientific content of each Antarctic expedition.

The Indian government's decision to launch an Antarctic expedition in 1981–2 was widely supported in the *Lok Sabha* and the national media. The unfurling of the Indian flag and the deposition of some inscriptions in English and Hindi by the leader of the first expedition attracted some complimentary commentary from the *Times of India*. Shortly after Gangotri I returned to India in January 1982, Gangotri II departed for Antarctica in February 1982 and Gangotri III had been accomplished by the season of 1983–4. The Gandhi government confirmed that at least $10 million would be devoted to polar expeditions and to the maintenance of the research stations. In the space of three years, the Indian government had completed three major polar expeditions and, "Its leaders are sufficiently assured of the country's overall industrial, scientific, and fiscal strength to announce a marked expansion of its Antarctic programme" (Larus 1984:55).

The annual reports, prepared by the Indian National Science Academy, to the Scientific Committee for Antarctic Research (SCAR) are testimony to that growth in scientific activity. Scientific research has ranged from biological investigations into the distribution of krill in the Southern Ocean to geophysical and geological research in the Schirmacher Range close to the Indian polar station (see *Indian Antarctic Research Report Number 1*, 1985). India's interests in non-living and living resources is similar to other southern hemispheric nations such as Argentina and South Africa. The official report of the First Indian Expedition noted that: "India should have a strong project on the exploitation and utilisation of krill resources . . . a study of the marine geology of the Antarctic continental margin will not only provide valuable information on the remit of the southern continents but also its mineral and oil potential which is reported to be promising" (Government of India 1982:60–3). The collection of weather data in Queen Maud Land and the investigation of the mineral potential of the Antarctic are often cited as the two most important strands of Indian Antarctic research. Whether or not understanding and the prediction of monsoon weather will actually improve as a consequence of this research remains to be documented. In fairness, Indian research on the consequences of global pollution has yielded some interesting insights into the movement of debris within the global atmosphere. The Gondwana theory of continental drift has been a popular geographical motif on which to hang resource and strategic debates. Although there remains much debate over the extent of the Antarctic's resource value, there has been some speculation that Indian research in Queen Maud Land had been attempting to assess the mineral potential of mountain ridges.

The international ramifications of India's involvement in the Antarctic remain of great interest. Government documents such as the *Seventh Five Year Plan* (1985–90) had clearly specified the intention of creating two permanently manned stations in the Antarctic before the possible review of the ATS in 1991 (Government of India 1985:374–5). This has now been partially achieved through the creation of the Maitri station and the summer station of Dakshin Gangotri. Admission to the international polar community had been granted swiftly, with India becoming a Consultative

Party member of the Antarctic Treaty System in September 1983 and the Scientific Committee of Antarctic Research in 1984. The Antarctic Treaty stipulated that accession to "consultative status" could only come about after an applicant had demonstrated "substantial scientific activity" (Article IX, para II), but it is significant to note that other states such as Poland and West Germany had to wait considerably longer than two years for Consultative Status.

The speed of the international recognition was no doubt related to the delicate timing of India's first Antarctic expedition, coinciding as it did with the conclusion of negotiations relating to UNCLOS III for the ocean floors and common heritage (Joyner 1992). The leading OECD nations had rejected the calls from India and the Developing World to discuss Antarctica as part of the "common heritage of mankind". Indian scholars such as Sanjay Chaturvedi and Anita Dey insist that in 1982–3 India was intent on persuading members of the international community that Antarctica should be considered a global common and henceforth placed under UN supervision rather than under the political and legal remit of a select international body such as the ATS (Chaturvedi 1994, Dey 1992). Other commentators suggested that India's formal relationship could have facilitated such a political transformation if that had been the case. It had been anticipated in the 1980s that India might undermine the authority of the ATS (which included South Africa), but in reality the relationship with the proponents of the common heritage was far more problematic.

India and the "Question of Antarctica"

India's commitment to the Antarctic remains noteworthy not only for the international politics of the region but also for Third World politics – with the added cachet of being the first nation outside the worlds either of advanced capitalist nations or the communist bloc to undertake research in the southern polar region. The Gangotri expeditions were undoubtedly pathbreaking in terms of consolidating the belief that India had both the resources and the technology to undertake such missions. Leaving aside the objections of cost and pressing social priorities closer to home (Dey 1992), the implications of these ventures were also recognized within global forums such as the United Nations and the Non-Aligned Movement. The first expedition coincided with an extraordinary growth of interest amongst Third World nations, especially Malaysia, as to the future management and possible exploitation of the Antarctic. It was seen by some commentators as offering opportunities to Third World states to criticize (on the inside as it were) and influence the economic exploitation plans of the First World as epitomized by the CRAMRA proceedings (Shusterich 1984, Chaturvedi 1990, Dey 1992).

India's attitude towards the Antarctic Treaty System has always been touched by a degree of ambivalence and contradiction. In the early years of the ATS, Arthur Lall, the former Indian representative to the United Nations, conceded that the Treaty covered many of the initial objections of the Indian government. The preamble to the 1959 Antarctic Treaty recognized that: "a treaty ensuring the use of Antarctica for peaceful purposes only and the continuance of international harmony in Antarctica will further the purposes and principles embodied in the Charter of the United Nations". India did not raise any formal objections after the ratification of the Treaty in June 1961, and indeed some commentators have suggested that the formal recognition of the authority of the United Nations (in tandem with the possibility of formal accession to the Treaty; see Article XIII) within the Treaty was sufficiently explicit to dampen Indian objections (see, for example, Beck 1986). But Lall later noted that in the 1970s the Treaty "tended to foster the idea that general international agreement can best be negotiated by a small group of countries, and in this respect the manner in which it was made is somewhat out of step with the requirements of an interdependent and shrinking world" (cited in Bhadauria 1983:575). Until the late 1970s, that form of objection to the ATS was generally muted because the "Question of Antarctica" was not considered a pressing concern for most of the Third World in the light of managing fragile economies, decolonization and the political consequences of the Cold War (Clapham 1985).

Indira Gandhi chaired the seventh meeting of the 101 non-aligned countries in March 1983 and, for the first time in the history of the NAM, the delegates included in their final report a resolution calling for a new universal framework of international co-operation under the authority of the United Nations. The NAM declaration in New Delhi demanded that the "exploration and exploitation of its [Antarctica] resources shall be carried out for the benefit of all mankind, and in a manner consistent with the protection of the environment in Antarctica" (cited in Mohan 1984:165). Resource difficulties too were now considered international political problems which touched upon profound economic divisions between North and South rather than East and West. In August 1983, the governments of Malaysia and Antigua-Barbuda requested that the secretary-general include the Antarctic on the agenda of the General Assembly under the heading of the "Question of Antarctica". This request was referred to the General Committee for consideration. Unsurprisingly, the members of the ATS were opposed to the motion and refused to participate in any ensuing discussion, but that opposition was not sufficient to prevent such an initiative becoming part of the agenda of the United Nations (see Beck 1986).

At the same time, India, as a major non-aligned nation, announced to the international community that she had signed the Antarctic Treaty and was hoping to join the ATS as a Consultative Party. In September 1983, the

Consultative Parties met in Canberra and approved India's request for admission. The political and diplomatic manoeuvres between March and September 1983 were numerous, and there is no question that India's formal involvement in the ATS was a considerable "blow" to the non-aligned group of nations intent on declaring the Antarctic a site of "common heritage" (Mohan 1984, Chaturvedi 1990, Dey 1992). Having supported the initial declaration of the 1983 NAM meeting held in New Delhi, the Gandhi government then wanted the international recognition of the ATS community. Indian officials probably reasoned that membership of the ATS combined with involvement of the NAM strengthened India's position to influence proceedings and dramatically change the international politics of Antarctica. As S.Z. Qasim noted to the 12th ATCP meeting in Canberra during September 1983:

> I thought it would be an appropriate occasion, Mr Chairman, to repeat what our Foreign Minister has said while informing the two Houses of Parliament after India's accession to the Antarctic Treaty and I quote: "By joining the Treaty, India would thus be able to exchange scientific information with other members of the Treaty and thereby enhance its analytical capabilities. India could hopefully be able to participate in the meetings of the Consultative Committee, and in doing so, will effectively project her own views as well as those of the non-aligned countries." (ATCP 1983)

The possibility of an Indian representative, inside the confines of the ATS structure, advocating a universalistic position on the Antarctic, did not materialize. The debates on the future management of the Antarctic, during the 1980s, witnessed a subtle change in the Indian position *vis-à-vis* the ATS. The Indian submission, in response to calls for the secretary-general to prepare a report regarding the Antarctic, was ambivalent in tone and substance. Whilst India had been critical of the secrecy in the ATS, the submission to the United Nations in 1984 now signalled a qualified support for the ATS:

> The Antarctic Treaty System . . . is an evolving institution whose structural and organisational frame-work is conceived in a flexible manner . . . which is gradually evolving further, taking into account the legitimate concerns of all . . . It would be unrealistic and counterproductive to think of a new regime in the present situation . . . The Antarctic is an open system . . . and should be broadened by the accession of more states. The evolving Treaty System should be made more open and responsive to the viewpoints of all states. (UNGA 1984)

This statement was effectively repeated by S.Z. Qasim in 1985 at the opening ceremony of the 13th ATCP meeting in Brussels, after welcoming China and Uruguay to Consultative Party status. Justifiably, Malaysia, who actively campaigned for the dismantling of the ATS, was not impressed by India's statements either on the ATS or Antarctica as a common heritage.

Thereafter, India robustly defended the ATS and its handling of the management of Antarctic affairs. For the "outsider", it appeared that India had maintained an impressive political and diplomatic balance and perhaps not suprisingly was later joined by other ATS members such as China in voting at the 40th UN General Assembly Session in November–December 1985 for the expulsion of South Africa from the ATS (Beck 1986). This in turn fractured the so-called "tradition of consensus" since a principal element in that "consensus" was the depoliticization of South African involvement in the Antarctic Treaty System (Dodds 1995). In contrast, Britain and the United States considered South Africa as just another nation actively interested in the Antarctic rather than an apartheid regime which had illegally occupied Namibia and destabilized most of southern Africa (Hanlon 1986, Sidaway and Simon 1993). Given India's longstanding objections to apartheid, voting in favour of a proposal for expulsion was perhaps inevitable and one that could not have been easily dismissed by an appeal for retaining consensus inside the ATS.

Generally speaking, India's strategy in the United Nations and elsewhere during the 1980s was consistently underwritten by concern for four issues:

As a littoral state, India strove to maintain the Indian Ocean and the Antarctic as a zone of peace (Hewitt 1992). In the absence of superpower competition, India had been better able to pursue its scientific, economic and resource priorities in the region. The growing size of her navy was interpreted by regional neighbours as a telling sign of India's intention to protect these varied interests in the region (E. Morris 1989). Likewise, the recent evolution of the Indian Ocean Rim Initiative illustrated India's commitment to develop economic co-operation amongst neighbouring oceanic states such as Australia, Kenya and South Africa.

The determination to play a part in shaping the creation of living and non-living resource regimes for the Antarctic was epitomized by India's decision to accede to the ATS in the midst of passionate debates over minerals. As the *Annual Report* of the Department of Ocean Development noted:

> Almost all the oceanographic features of the Indian Ocean, including its fertility, are governed by the Southern Ocean. The need for data on the biomass and population density of krill, and of seabirds and marine mammals are [*sic*] essential to assess the potential and sustainable yield of the exploitable living resources from the northern sector [Indian Ocean] to the southern sector [Southern Ocean]. (DOD 1988:33)

India's national prestige has now been symbolically linked to Antarctic programmes and, as with other nations, whether in the northern or southern hemisphere, activity in the Antarctic continent signifies the scientific and social strength of a nation. Indian delegations to the ATCP conferences in the 1980s consistently maintained that active participation of Third World states in the ATS could help to achieve the political objectives of NAM rather than lead to their dissolution.

India, resource regimes and the Protocol on Environmental Protection

Clearly affected by the UNCLOS III negotiations in the 1980s, India, armed with a budget of $30 million devoted to deep seabed mining ventures, initiated exploration of 1.5 million square kilometres of the central Indian Ocean (Government of India, MIB 1993:170–1). Whilst anxious to be declared a pioneer investor by the Preparatory Commission of the International Seabed Authority, the central role of the Department of Ocean Development in formulating policies relating to resource and territorial claims should not be underestimated as, according to Dr S.Z. Qasim, the secretary of the DOD in 1983: "India's accession [to the Antarctic Treaty] did not imply any change in its basic stand either regarding territorial claims [India rejected all claims to the Antarctic] or regarding the treatment of the natural resources of Antarctica as the common heritage of mankind" (cited in Dey 1992:178). However, the government of India has repeatedly confirmed that scientific programmes on the Antarctic are linked to their concerns for the oceanography of the Indian Ocean and to deep seabed mining.

In order to participate in the ATS discussions for living resource management, India acceded to the Convention on the Conservation of Marine Antarctic Resources (CCAMLR) in July 1985 and gained admission as a formal member of the Commission in June 1986. The involvement in the management and possible distribution of living resources in the Southern Ocean remains vital given persistent resource worries in New Delhi: "To the undernourished, protein starved masses of India, about to enter the 21st century with over a billion mark, Antarctica's marine resources could be a welcome food . . . the fourth Indian expedition mounted a large scale operation to study the possibilities of exploiting krill from the southern waters" (Chaturvedi 1990:166). Consequently, Indian delegations to the CCAMLR negotiations and the ATCP meetings have consistently argued for greater research into the resource potential of the Southern Ocean whilst calling for the introduction of effective measures of resource conservation.

From June 1982 onwards, the ATS again attempted to resolve the issue of a mineral resource regime for the Antarctic, originally discussed at the ATCP Wellington conference of 1970. Whilst Malaysia called for an international moratorium on the question of Antarctic resources, the ATS tried to negotiate a new convention. India participated in these minerals negotiations from January 1984 until the eventual agreement of the Protocol in Madrid in June 1991. British scholars such as Peter Beck voiced concerns. "Of course, it might be argued that India intends to exploit entry [into the ATS] in order to destroy the Antarctic club from within, or to exacerbate the existing internal divisions, only time will tell" (Beck 1983:299). The debates over a minerals regime were the most difficult in the history of the ATS, culminating at a time when the system was subjected to intense political pressure from the non-aligned nations and the United Nations (Parsons 1987). As a major representative of the non-aligned world, India was not expected to be a strong supporter of a new minerals regime which would have been widely interpreted within the Group of 77 as serving the interests of the Developing World. By the mid-1980s, however, India was aided by the fact that a number of other nations such as Uruguay, China and Brazil could share the mantle of representing the interests of the Developing World inside the ATS. A part of the problem was plainly that the position of the Third World towards the Antarctic was assumed by the Group of 77 to be monolithic rather than fractured. This was compounded by the attitudes of long-standing members of the ATS such as Argentina and Chile, who, whilst expressing support for the general principle of common heritage, were hostile to such an application for the Antarctic.

The Convention on the Regulation of Antarctic Mineral Resource Activities (CRAMRA) was completed in June 1988 after six years of discussion and often tortuous negotiations (Beeby 1988). The final act was signed by six developing countries (including Argentina, Chile, South Africa and India) and 14 other nations including Australia, New Zealand and the United Kingdom. The ratification of CRAMRA depended upon the signatures of all seven claimant nations, the superpowers and at least seven non-claimants such as India or South Africa. CRAMRA proposed a regulatory framework for the possible stages of mineral development. Geographical regulation was extended to the continent, islands and adjacent offshore areas south of 60 degrees South latitude but not to the deep seabed (see Joyner 1992). The Indian government considered it to be a sensible mechanism for the regulation of polar prospecting, exploration and development rather than a mining code for inevitable exploitation. To its opponents, however, CRAMRA was a mining convention and it had to be resisted with vigour. In May 1989, the governments of France and Australia reversed their earlier agreement to CRAMRA and called for a new regime based on the concept of a world wilderness and a permanent ban on mining.

India shifted its support of CRAMRA in 1990–1 to that of the Franco-Australian position on the need for environmental protection and a mining moratorium (Sreenivasan 1994). The CRAMRA proceedings were undoubtedly imperilled, as was the ATS, by this change in policy direction by two of its leading proponents. Whether the growing strength of green politics in France and Australia or the activities of international pressure groups such as Greenpeace were instrumental in prompting this policy shift remains to be clearly demonstrated (Bergin 1991, Darby 1994). There is no doubt that the ATS split over the issue of mineral management, as nations such as Belgium, New Zealand and Sweden supported the overturning of

CRAMRA, whilst the British, Argentine, Chilean and American delegations fought for its retention. India's support for a temporary ban on mining was reflected in its acceptance of the 1991 Protocol on Environmental Protection. The Protocol itself designated the Antarctic as a "natural reserve devoted to peace and science", and Article 7 categorically prohibited any activity relating to mineral resources. As one official Indian publication generously concluded: "A new protocol on the protection of the Antarctic which included a prohibition of commercial mining in Antarctica was agreed to for which India worked relentlessly" (Government of India, MIB 1993:166).

India's role in the international politics of the Antarctic will continue to be shaped by the competing demands of national interests, the international community and regional concerns in South Asia. It remains to be seen whether India can perform the role that some Indian scholars outlined in the midst of the mineral controversies: "India would have to play a crucial role in order to safeguard the interests of the developing world – to ensure that the environment is protected and to prevent any resource grab by a few states or corporations" (Mohan 1984:167). The ratification of the Protocol by the Indian government is probably a reflection of the fact that Indian interests in the possible resource exploitation of the Antarctic region have not been compromised there. Moreover, given that this temporary ban will have to be renegotiated in the next 40 years, Indian political leaders may well feel that they will be in a better position, in terms of technological development and practical experience of ocean mining, to maximise any future opportunities for resource exploitation.

The interaction between resources, scientific investigation and oceanographic research is particularly profound in the Indian case. As the objectives of the 8th plan of the Department of Ocean Development reiterated: "to improve the surveillance, management and the exploitation of marine resources; to safeguard the long term interests of the global commons; and to develop marine infrastructure in India" (Government of India, MIB 1993:165). The creation of an Antarctic Study Centre at Vasco in 1992 and the upgrading of oceanographic vessels *Sagar Kanya* and *Sagar Sampada* must be considered indicative of these stated aims. From being a sceptical commentator on the international politics of the ATS, India has emerged in the 1990s as a major Third World Consultative Party member. As the 10th largest economy in the world, India has been able to devote considerable resources to undertake and develop a logistical and scientific polar programme which is intimately linked to oceanographic and resource research in the Indian Ocean. By 1995, 14 expeditions had been dispatched to the Antarctic and to the Weddell Sea. A rising population coupled with growing resource demands will ensure that India retains an active interest in the resource regulation of the Antarctic and in the evolution of UNCLOS derived deep seabed mining rights in the Indian Ocean (Government of India, DOD 1995).

Summary

India's interest in the international politics of the Antarctic and the Southern Ocean spans half a century. The earliest calls for the United Nations General Assembly in the 1950s to consider the political and legal status of the polar continent sent shockwaves through the capitals of Britain, the old Commonwealth and South American states. Whilst most Indian attention has been focused on the immediate areas of the Indian Ocean, the importance of India to international politics of the Antarctic should not be underestimated. In the space of two decades, India established herself as a major so-called "developing nation", with active interests in the Antarctic. The accession to the Antarctic Treaty System in 1983 confirmed in part the fears of the international community, that large regional powers such as India and China would have to be included within the ATS. In contrast to the earlier accession of West Germany, India was admitted on the basis of planned scientific research in the Antarctic rather than on the limited achievements of her first three polar expeditions. Of immense concern to India in the near future will be the attitude of non-signatory Pakistan to the ATS. As a sometime supporter of the Malaysian challenge in the UN, Pakistan's expeditions to the Antarctic were not subject to the obligations and commitments of the ATS. Whilst there have been vague indications from the Pakistani Department of Foreign Affairs that it will be reviewing the previous policy of hostility to the ATS (as witnessed by its election to SCAR in June 1992 as an Associate Member), future membership of the ATS may ultimately depend on the attitude of an Indian government. Moreover, the ATCPs will also have to decide whether Consultative Party status should be awarded to Pakistan immediately, as the latter expects. A further issue to be considered will be India's ability to play a role in shaping the conclusions of the Transitional Environmental Working Group on the management of the liability annexe, given its interests in the resource and marine management of the southern part of the Indian Ocean.

7

New Zealand and the Ross Dependency

New Zealand's role in the Antarctic is paradoxical and perhaps the most interesting of the four claimant SORS. Francis Auburn argued: "The New Zealand government took no active interest in the [Ross] Dependency from 1925 to 1955" (Auburn 1972:57). In marked contrast to their antipodean neighbour, New Zealand did not dispatch a single national expedition to the Ross Dependency. Active involvement, during this period, was minimal with the exception of the participation of Sir Edmund Hillary in the 1955–8 Trans-Antarctic Expedition (TAE). In 1955, the New Zealand government offered to extend all facilities to the US expedition Deep Freeze and gave it permission to establish its own postal services "in the absence of ordinary postal facilities in the Dependency" (cited in Auburn 1972:57). In the same period, Prime Minister Walter Nash raised the possibility of internationalizing the Antarctic via the United Nations and the formal abdication of her sovereign claims to the Ross Dependency (see Map 7). Whilst maintaining a claim to the Antarctic, New Zealand has been a committed contributor to the international politics of the Antarctic Treaty System and her diplomats have: "never hesitated to push for changes to deal with what it regards as unfinished and incomplete formulations in the original treaty" (Alley 1984:911).

New Zealand's historical association with the Antarctic region was shaped by her connection with the British Empire. In spite of the political ambivalence towards the Ross Dependency claim, it was deemed that relative proximity to the Antarctic compelled New Zealand to retain an interest in the political and resource management of the polar continent and the Southern Ocean. The close logistical and political relationship with the United States during the Cold War was reflected in New Zealand's decision to co-ordinate over polar logistics. In conjunction with Australia,

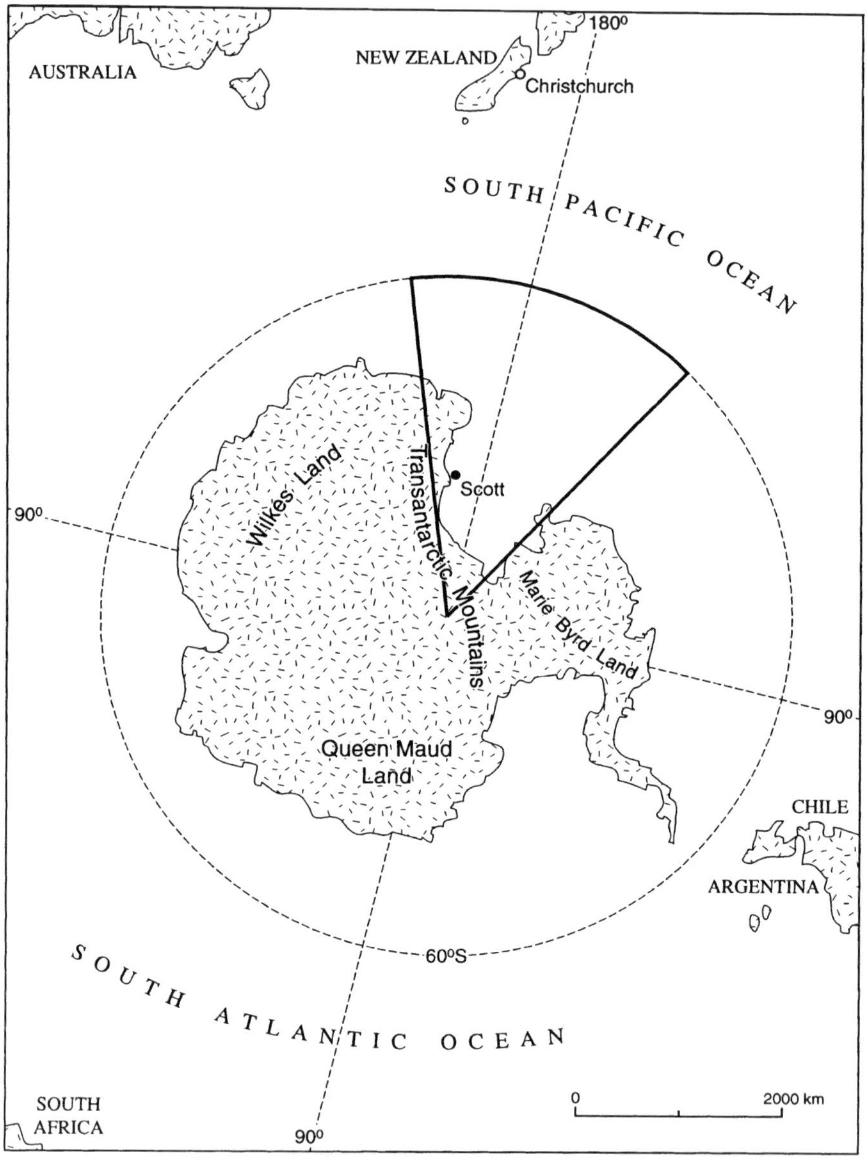

Map 7 Antarctic territory claimed by New Zealand, and New Zealand's permanent base on the continent

successive post-war governments in Wellington have also cemented regional security alliances with the United States. Later, New Zealand's delegation to the Antarctic Treaty Consultative Meetings played a key role under the chairmanship of Chris Beeby during the protracted negotiations over CRAMRA (1982–8) and the subsequent negotiations over the Protocol on Environmental Protection. During the 1980s, therefore, New Zealand and Australia (as chair of the UN ATCP group) performed significant roles in terms of securing a public consensus over resource management and the international organization of Antarctic politics. New Zealand has also been the most willing of all the SORS to consider relinquishing her territorial sovereignty in the Ross Dependency.

This chapter initially considers the earliest phase of New Zealand's activities in the region and her relationship with Britain which in part overlap with the experiences of Australia and South Africa. The actions of successive governments in the inter-war period are explored, together with New Zealand's ambivalence over the resource management of the Ross Sea region and the exploratory activities of other states such as the USA and Norway. An appraisal of the gradual re-emergence of New Zealand's polar interests in the immediate post-war period is then considered. New Zealand's participation in the 1959 Washington Conference is analysed together with her assessment of the Antarctic Treaty. The inter-relationship between New Zealand's Antarctic interests and those in the South Pacific are then examined with a view to articulating the regional agendas implicit within New Zealand geopolitical reasoning. Her role in resource management, especially during the CRAMRA negotiations, is investigated because New Zealand played a major role in the ATS, in her capacity as chair, at the minerals regime meetings. The attitude towards the Protocol on Environmental Protection is also considered. Finally, some brief conclusions on the geopolitical significance of New Zealand's involvement in the Antarctic and Southern Ocean are noted in the light of recent institutional changes resulting in the creation of a new agency, Antarctica New Zealand.

New Zealand and the Ross Dependency (1923)

At the turn of this century, Christchurch and Dunedin established themselves as staging posts for British expeditions to the polar continent. The New Zealand government had offered financial support to the Scott and Shackleton expeditions, and local reaction to these developments tended to be supportive of British imperial endeavours. The *Lyttelton Times* noted in October 1910 that: "if the Union Jack is flown over the South Pole there will be no doubt in Lyttelton's mind that the honour and the glory belong to her as much as anybody", and again in November 1910 that "the enterprise that the explorers will try to complete is one very dear to the whole Empire" (cited in Logan 1979:20). There were, however, other occasions when the New Zealand government was not supportive of polar expeditions such as the 1911–14 Australasian Antarctic Expedition led by

Douglas Mawson. In contrast to their continental neighbour, New Zealand was not an active participant in the Antarctic region.

During 1919–20 on the basis of a number of interlinked arguments, the British government approached the Commonwealth government of New Zealand over the possible administration of the Ross Sea. The first meeting between the Colonial Office and the representatives from Australia (Senator Millen) and New Zealand (Sir James Allen) took place in February 1921. The official document, *Memorandum on Control of Antarctica*, produced after the meeting stipulated that Australia and New Zealand should have separate territorial responsibilities divided at 160 degrees East. The document was one of the earliest attempts by a state to stipulate the conditions for a national claim bar a few existing precedents. The problem facing British administrations in the 1920s was that their claims to have discovered vast chunks of the Antarctic were either contestable in their own right or contested by other explorers and their nations. The Norwegian explorer Roald Amundsen had visited the regions considered relevant by British officials. There had also been French, American, German and other European expeditions as the secret papers and maps of the Colonial and Foreign Office acknowledged. The memorandum argued that for initiating further annexation, the New Zealand area was probably the most accessible part of the Antarctic after the peninsula region. British officials, in conjunction with individuals such as Frank Debenham of the Scott Polar Research Institute, urged the New Zealand government to issue an Order-in-Council in case the Norwegians pressed either a counter-claim or over-exploited the whaling stock of the Ross Sea (Auburn 1972, Logan 1979). In July 1923, using the precedent of the British Settlement Act, the Massey government issued an Order-in-Council which declared that the Ross Dependency would consist of the coasts of the Ross Sea along with "the islands and territories adjacent thereto between 160 degrees of East longitude and the 150th degree of West longitude, which are situated south of the 60th degree of South latitude" (cited in Bush 1982). These regions were declared a British settlement and were to be administered as a British territory under the authority of the governor-general of New Zealand.

New Zealand's foreign policy during the inter-war period was primarily shaped by the demands of Britain. The Department of External Affairs had minimal responsibilities and was largely concerned either with managing regional relations or with liaison with London (Logan 1979). However, the proposed activities of the Rosshavet Whaling Company in the 1920s had raised difficult issues in terms of effective management of the region and Anglo-Norwegian rivalry over possible claims to the Antarctic continent and surrounding islands. The Norwegian company's application to catch whales in the Ross Sea had forced the New Zealand government to decide on a strategy for the regulation and management of living resources.

The eventual application to Rosshavet set an important precedent in terms of administrative control.

During the 1926 Imperial Conference, the British approached the New Zealand Prime Minister Gordon Coates with the suggestion that the Ross Dependency might be extended eastwards towards the French claim of Adélie Land. But New Zealand government records note that they were opposed to any extension of the Ross Dependency and were more concerned with the question of Ross Sea whaling and effective resource regulation (Logan 1979:39). New Zealand's vision of the Antarctic was significantly different, therefore, from British representations of continental control (Quartermain 1967). The authorities' geopolitical vision of the region was based on effective resource regulation and mutual co-operation between states. The development of pelagic whaling and Norwegian insistence that New Zealand's maritime jurisdiction was limited to a three-mile zone made land-based control of whaling harder to achieve. The solution to these administrative dilemmas was to introduce a system of licensed operators and a set of whaling regulations for the Ross Sea. The period between 1926 and 1930 was further complicated by the fact that the Norwegian government had encouraged Lars Christensen to annex territories east of the Ross Dependency, including the disputed island of Bouvet.

New Zealand's willingness to contemplate international co-operation and multilateral relations were based on a sober appreciation of the international politics of the Antarctic in the late 1920s and 1930s. Influencing this vision was the growing involvement of Norway and the United States in the region. The expeditions of Richard Byrd had immediate relevance because of concerns that Byrd's flights over territory east of the Ross Dependency were the prelude to American territorial claims (Dodds 1997). American ambivalence to polar claims was gradually reversed largely as a result of the endeavours of Byrd and his staking of claims in Marie Byrd Land, east of the Ross Dependency. The fiction of administrative control was to become more apparent as the 1930 Imperial Conference approached. The BANZARE expedition coincided with renewed Norwegian interest in the Antarctic continent. The ship *Norvegia* had been dispatched to the region at the same time as New Zealand issued new whaling instructions for the Ross Dependency. Fears of a diplomatic incident and physical clash in the Antarctic were averted when the BANZARE expedition took their ship *Discovery* east and the Norwegian expedition went west of Enderby Land. It was quite clear that these incidents involving not only the United States but also Norway had effectively compromised British plans for complete annexation of the continent. The new priorities of the Imperial Conference were, therefore, to be the establishment of Australian territorial claims and the effective maintenance of claims in the FID and the Ross Dependency.

The Deterritorialization of responsibility (1930–58)

The period between 1930 and 1958 was characterized by scant official interest in the Ross Dependency. Indeed, the minimal participation of New Zealand in the BANZARE had inspired the creation of the New Zealand Antarctic Society in 1933. This private organization was effectively responsible for the promotion of interest in the region as the government made "little attempt . . . at even the nominal administration of the New Zealand claimed Antarctic territory" (Hanessian 1965:19). The New Zealand Antarctic Society had, therefore, three major objectives: to promote interest in the Antarctic within New Zealand; to encourage other expeditions to come through New Zealand and to encourage the creation of a New Zealand research programme for the region. The retirement of political figures such as Dillon Bell and Walter Massey coupled with the onset of economic depression in the 1930s are often held to be the chief causes of this decline in official interest (Logan 1979:70). However, growing tensions over territorial claims in the Antarctic persuaded New Zealand officials that it should remain interested in proceedings but as an onlooker rather than an active participant. The difficulties over Commonwealth Antarctic claims were typified by the experiences of the Australian government in the 1930s. The formal creation of Australian Antarctic Territory in 1933 effectively encompassed all of the continent between the Ross Dependency and 45 degrees East longitude, excluding the French claim. To the irritation of the Norwegians, the Australian claim had encompassed areas of Norwegian interest. The British claim to the FID was also being challenged by Argentina and Chile who had an active interest in the region. The Argentines had maintained a meteorological station in the South Orkneys since 1904 and in 1927 had informed the International Postal Bureau of their rights to the FID region (Fraga 1979).

With rising tension in the Antarctic region, the New Zealand government was content to offer assistance to the American expeditions, and the British supported this decision as officials in the Dominion Office believed that it would demonstrate the administrative polar capacity of the New Zealand government. The British government were concerned that the Byrd expedition might be the prelude to a formal declaration of American interest in the Ross region. This fear appeared confirmed when it was declared that Byrd had plans to construct a post office in Little America II (Klotz 1990:17). It was becoming increasingly clear that the claim to the Ross Dependency was not going to be recognized by the United States (Klotz 1990). According to one leading New Zealand legal expert, the claim to the Ross Dependency was increasingly acknowledged to have been based on a flawed application of the 1887 British Settlements Act and that the title to the Ross Dependency was "sadly defective" (Auburn 1972:64–6 and 71).

New Zealand's position on territorial claims in the Antarctic contrasts markedly with the Australian case and the sustained efforts of individuals such as Richard Casey and Douglas Mawson to locate the Antarctic in the Australian public consciousness in the 1930s. The 1937 Imperial Conference had recognized that the Commonwealth Antarctic faced further competition from Argentina, France, Norway and the United States. The situation was complicated still further by new developments: France altered its claim to the Adélie Land from its coastal limits to create a new sectoral claim which was extended between 136 degrees East and 142 degrees East longitude; the flying exploits of American and Australian aviators had captured the imaginations of officials and explorers in the SORS. The possibility of air routes over the South Pole called for new strategic assessments of the continent. In 1938, France and the Commonwealth nations signed a treaty of reciprocity over flight rights which effectively recognized the legitimacy of existing claims and apportioned legal rights over the polar airspace. The expanding whaling fleets of Germany and Japan were active in Antarctic waters and were not inclined to recognize the pleas of the 1931 International Whaling Convention to reduce total catch size. The Hitler regime decided in November 1938 to despatch the ship *Schwabenland* in order to operate between the Australian claim (AAT) and the FID. The ship's aircraft carried out aerial mapping and dropped territorial markers in the region previously explored by the Norwegian ship *Norvegia*. Eventually, in January 1939, the Norwegian government, concerned over the rising interest of the German expedition, issued a proclamation that a Norwegian territorial sector existed between 45 degrees East and 20 degrees East, completing the space between AAT and the FID. No northern or southern limits of the Norwegian polar sector were specified. Thereafter, the Antarctic boundaries of Britain, Norway and Australia were mutually recognized.

The attitude of the New Zealand government to these international endeavours was muted given that most of the activities were located outside the Ross Dependency. Of most concern to officials in Wellington were the ongoing developments of US Antarctic policy and exploration in Marie Byrd Land. According to observers, the attitude of the State Department in Washington changed when President Roosevelt ordered a re-examination of American policy in 1938 and confirmed to explorers such as Lincoln Ellsworth that it would be appropriate for an individual to make claims on behalf of the United States (Bush 1988:438). This growing interest of the US president in the Antarctic region was indicative of Roosevelt's conviction that circumpolar air routes had transformed the strategic significance of the Antarctic and the South East Pacific. Reports later indicated that President Roosevelt became interested in American rights in the Ross Dependency as a consequence of a US–New Zealand dispute over the ownership of the Phoenix Islands in the South Pacific (cited in Logan 1979).

The onset of the Second World War affected the international politics of Antarctica and New Zealand's involvement in the Southern Ocean. Whilst most of the claimant states and the United States were preoccupied with the war in Europe and the Far East, the territorial politics of the region were complicated by the formal polar claims of Argentina and Chile. In addition, German sea raiders had occasionally visited the sub-Antarctic island of Kerguelen and were using it as a major refitting point. Their ships had been highly effective in sinking Allied shipping, mining antipodean ports and capturing a portion of the whaling fleet of Norway. From the New Zealand perspective, the activities of the German navy had the effect of reterritorializing their sense of responsibility in the Antarctic and Southern Ocean. In contrast to earlier assessments, which tended to stress resource management and territorial indifference, there was a shift towards a more explicitly geopolitical evaluation of the Southern Ocean-based sea lanes, resource rights and territorial polar occupation. In 1941, New Zealand created observation posts on the island groups of Auckland and Campbell in response to fears that they were already being used by German forces.

In the aftermath of the Second World War, the Fraser government ordered a major revaluation of New Zealand's Antarctic and Southern Oceanic interests. The mechanisms for the effective administration of the Ross Dependency were considered. A memorandum from the Dominions Office to the Ministry of External Affairs on 13 September 1945 suggested that New Zealand would have to reconsider the legal basis of the Ross Dependency. The post-war revival in Antarctic exploration and territorial claims created a new context for New Zealand foreign policy. A telegram from the Dominions Office in December 1946 confirmed the implications of this shift: "in the light of the forthcoming United States expeditions to Antarctica, Commonwealth and New Zealand Governments would be well advised to take early steps to strengthen their claims . . . this could be effectively achieved by the establishment of at least one permanent base in each sector" (cited in Logan 1979). The effective living resource management of the Ross Sea region also remained a policy priority for the government in spite of the damage inflicted to allied whaling fleets during the Second World War. After considerable discussion in government circles, the final report of an Inter-Departmental Committee on Antarctica concluded in May 1947 that a small expedition might be economically justifiable rather than a larger programme of expeditions and base constructions. This modest proposal was in direct contrast to the Australian and South African governments, which had established bases on a number of sub-Antarctic islands (Heard, Marion and Macquarie) in response to the British government's agreement to transfer their sovereign rights in the region (Swan 1961, Cooper and Headland 1991). The Fraser administration did not share the strategic vision of individuals such as President Jan Smuts of South Africa or Robert Menzies of Australia. Fraser's deputy, Walter

Nash, later implicitly rejected such geopolitical arguments in favour of the promotion of international co-operation over the region.

The earliest arguments for the internationalization of the Antarctic are to be found in the records of the Fraser administration. The timing of these sentiments was also a reflection of wider circumstances that were affecting the international politics of the Antarctic such as the intensification of the Anglo-Argentine–Chilean disputes over the ownership of the Antarctic peninsula and surrounding islands. American attitudes towards the Antarctic were also conditioned by differing sets of relationships on the one hand, the evolving, if disputed, "special relationship" with the United Kingdom and on the other hand, the promotion of pan-Americanism and the concept of inter-American mutual defence (Child 1979). With the start of the Cold War, the US State Department had begun to shift away from the active promotion of US claims *per se* to the consideration of a new international regime for the governance of the Antarctic. The New Zealand government, whilst sympathetic to the notion of some sort of internationalization, nevertheless was also influenced by British objections to the idea of a United Nations solution. The British preference for a multinational condominium based on the existing claimant states and the United States would have, in theory, prevented the formal involvement of the Soviet Union. The Department of External Affairs in Wellington had expressed a willingness to explore the options, even though senior members of the Fraser administration preferred a solution constructed in the United Nations.

The desire to witness this internationalization of Antarctica was more complicated than was generally assumed within New Zealand. Public and official interest in the Antarctic region had grown in the late 1940s, as shown by the revival of the New Zealand Antarctic Society and their publication, *Antarctic News Bulletin*. Another indication of renewed territorial interest in the Ross Dependency came from the personal interest of Premier Peter Fraser in 1949. In response to a memorandum from the secretary of state for Commonwealth Relations, Fraser acknowledged interest in a proposed expedition to the Ross Dependency to be led by the director of the Scott Polar Research Institute, Frank Debenham. The expedition would undertake a detailed study of the ice mechanics of the McMurdo Sound on the eastern edge of the Ross Dependency (New Zealand Archives, Department of External Affairs 1949 cited in Logan 1979). According to the records of the Ministry of External Affairs, the Prime Minister approved the plan and hoped that New Zealand's territorial interests would be enhanced in the face of recent American expeditions (Hanessian 1965:18). In 1952, the New Zealand Antarctic Society produced an important book entitled *The Antarctic Today* which reviewed the current state of the Antarctic and carried contributions from other southern hemispheric nations such as Australia and South Africa. In conjunction with the editor of the *Antarctic News Bulletin*, L.B. Quartermain, the Society was active in promoting the notion that New Zealand had a geographical responsibility for the Antarctic region. In the final chapter of the book, Quartermain states:

> New Zealand can no longer be regarded as merely the "gateway to the south", a port of call for expeditions to the Antarctic. The New Zealand people are, in fact, part owners of Antarctica . . . The Ross Dependency is therefore New Zealand's particular responsibility . . . It has long been recognised that claims to sovereignty must rest on something more tangible than mere geographical proximity . . . To establish recognised sovereignty there must be effective permanent occupation . . . But New Zealand – the administrator of the Ross Dependency – has gone no further than to occupy Campbell Island, in lat. 53 degrees South, as a permanent weather station . . . The time must surely have come for New Zealand to establish effective occupancy of that sector of the Antarctic which is peculiarly hers, and for adventurous and qualified young New Zealanders to be given the opportunity to study their portion of the Antarctic for themselves. (Quartermain 1952:365)

In spite of these "geographical" appeals, the new National government of Sidney Holland, elected in late 1949, remained intent on matching economic means to political priorities and the Ross Dependency was not considered a geopolitical priority. The *Evening Post* of Wellington noted in August 1950 that the minister of external affairs had acknowledged New Zealand's strategic interest in the region despite the lack of "scientific discovery and exploration". This public clamouring for greater involvement in the Antarctic, however, increasingly clashed with the principles of internationalization held by politicians such as Nash (Dodds 1996b).

The role of New Zealand in the Antarctic and the Southern Ocean in the years leading up to the Antarctic Treaty was paradoxical and perhaps even confused (Simpson 1954, Quatermain 1967, Peat 1983). New Zealand was persuaded to participate in events such as the Trans-Antarctic Expedition which, in spite of their scientific pretensions, were designed to promote Commonwealth territorial claims. Yet political leaders such as Walter Nash suggested, along with other nations such as Brazil and India in the 1950s, that the Antarctic should be freed from territorial claims and placed under the authority of the United Nations. The impetus for the rising involvement in the Antarctic in the 1950s was international political pressure from Commonwealth states such as Australia and Britain coupled with the domestic campaigns of the New Zealand Antarctic Society led by the experienced explorer and former president of the New Zealand Royal Society, Dr R.A. Falla.

The most significant moment in New Zealand's renewed engagement with the Antarctic came with the announcement that the governor of the Falkland Islands, Sir Miles Clifford, had formally proposed to the Polar Committee in London the undertaking of a Trans-Antarctic crossing in March 1953. The author of the original proposal was the leader of the Falkland Islands Dependency Survey, Dr Vivian Fuchs, who had already collaborated with the president of the Royal Geographical Society, James Wordie (Fuchs and Hilary 1958). In September 1953, Fuchs outlined his plan to the Polar Committee with its mixture of representatives from British governmental departments and the Commonwealth nations of Australia, Canada, South Africa and New Zealand. The proposal was conceived as a Commonwealth project designed to promote international political and scientific collaboration as well as national prestige (Dodds 1997). The political value of the expedition was derived from the planned route which would cross the territories of the FID, the Australian Antarctic Territory and the Ross Dependency. Even though the Polar Committee minutes of 15 September 1953 acknowledged that the territorial rights might not be affected by the venture, most of the people involved believed that the expedition would be beneficial in view of South American and other territorial competition (Dodds 1996b).

In considering the interests of New Zealand, the Churchill government proposed that the TAE might lead to the creation of a base on Ross Island thereby further protecting Commonwealth territorial claims. As with the Canadian government's reactions to the Fuchs project, the New Zealand government was not prepared to offer a recommendation on the transcontinental crossing in September 1954. Ironically, the Governor-General of the Ross Dependency had informed the Ministry of Marine in June that year that the government had two alternatives – either renounce their claim to the region or establish an effective presence (Logan 1979:122). The announcement of the organizing committee of the IGY (the Comité Speciale de l'Anée Geophysique Internationale) in Rome was a catalyst for action as it requested that the New Zealand government create an IGY station between Ross Island and Cape Adare. The New Zealand government was, therefore, effectively pressurized into contributing to the TAE and in January 1956 the Ministry of External Affairs agreed to consider participation in the TAE and IGY programmes. At a Commonwealth meeting in London on 1 February 1956, the Churchill government announced that £100,000 would be offered to the TAE in conjunction with £25,000 from Australia and £10,000 from a reluctant South Africa. In spite of New Zealand's misgivings, New Zealand later approved £25,000 for the TAE and agreed to investigate participation in the IGY. The decision of the cabinet was announced to the press on 15 May 1955 and confirmed that the government had, amongst other things, created a Ross Sea Committee to oversee the construction of a reception base in McMurdo Sound (Fuchs and Hillary 1958).

Despite the TAE's publicity, most government ministers, whether in Wellington, Pretoria or Canberra recognized that the IGY would probably be significant in the political and scientific long term. The importance of this new commitment was dramatically emphasized in June 1955 when the United States announced that she intended to create a logistical base in

McMurdo Sound (Klotz 1990:28). This decision confirmed that the American expedition *Deep Freeze* would be operating for the first time at the western edge of the Ross Dependency. The Department of External Affairs acknowledged that American activity in the Ross Dependency should be subject to a diplomatic note (sent on 26 June 1955) acknowledging New Zealand's administrative responsibilities (New Zealand Archives, Department of External Affairs 1955 cited in Dodds 1996b). The response from Washington was to propose a joint New Zealand–US base in McMurdo Sound rather than to reject the New Zealand claim outright. This proposal was accepted by the New Zealand government who in the process initiated a relationship, which still exists today, of American logistical support for New Zealand's activities in the Ross Dependency. At the time, the Ministry of External Affairs was also concerned that the Japanese might locate a base in the region and hoped that this joint proposal would dissuade the government in Tokyo.

The creation of the Scott base in the Ross Dependency, whilst politically and symbolically astute, was a lengthy affair. With the purchase of the supply ship HMNZS *Endeavour* from the British government in 1955–6, the New Zealand government gradually assumed a more prominent role in the TAE. In terms of administrative niceties, Sir Edmund Hillary was appointed a magistrate and postmaster for the duration of the TAE. In return for an investment of more than £250,000, it was agreed that on completion of the TAE, New Zealand would take over the Scott base thereby committing the government to an Antarctic presence. The geopolitical reasoning employed by the New Zealand government to explain its involvement in the TAE echoed that of the Menzies government in Australia: strategic interests, geographical proximity and resource potential. There was one significant difference: anti-communism or stated fears over the intentions of the USSR were less intense (Cunningham 1958).

The national election in 1957 coincided with a series of interesting developments affecting New Zealand and the Antarctic. In spite of the presence of the Scott base, the Department of External Affairs pressurized the Prime Minister into maintaining a commitment in the polar region and the claim to the Ross Dependency. In the midst of the Cold War, the IGY witnessed the Soviet Union establishing a base in the AAT, Britain undergoing territorial disputes with South American states, and Japan and South Africa raising the profile of their interests in the polar region. At the same time, proposals for the internationalization of the Antarctic were finding some favour in Wellington with the Labour leader Walter Nash a firm advocate of this kind of regime. Prior to the Washington Conference, other Commonwealth leaders were increasingly anxious that New Zealand's possible objection might restrict interested parties of the claimant states and those who had participated in the IGY polar programme. As a New

Zealand government press release confirmed in January 1958: "Mr Nash felt that some form of international regime might be a most satisfactory solution. He believed that no such regime could be successful which did not have the blessing of the United Nations. Such a solution would be in keeping with 20th century thinking and the objectives of the United Nations charter" (cited in Dodds 1996b).

Ironically, the decision of the Eisenhower government to seek a conference of interested Antarctic parties in May 1958 was greeted with enthusiasm by Prime Minister Nash. In contrast to the Australian government, the Labour premier argued, in a report published by the *Evening Post* (29 May 1959), that: "such a conference could well lead to an agreement on Antarctica which could open the way to international co-operation in other and more controversial areas". Nash's stance on international co-operation was politically advantageous for it could have removed the need to protect the rights of New Zealand over the Ross Dependency and, with logistical support from the United States, the expense of an Antarctic programme would be significantly reduced. Indeed, as the Ministry of External Affairs legal adviser Christopher Beeby later noted, this dependence on the United States was formalized by an exchange of notes in 1958 (Beeby 1972:9). Nash's idealism was subjected to fierce criticism from the New Zealand Antarctic Society, which was concerned along with leading national newspapers that Nash might compromise New Zealand's rights to the region before any new legal regime had come into existence. However, Nash had agreed to the creation of a new administrative structure to manage New Zealand's Antarctic affairs. In April 1959 the Antarctic Division of the Department of Scientific and Industrial Research (DSIR) was established to manage Antarctic affairs in conjunction with the Ross Dependency Research Committee, which drew upon the expertise of universities, the Royal Society and various government departments (Quartermain 1971).

New Zealand and the 1959 Antarctic Treaty

In the opening statement to the Washington Conference on 15 October 1959, the New Zealand representative A.D. McIntosh argued that:

> New Zealand has, in recent years, advocated international action to ensure the future use of Antarctica for the welfare of all nations . . . (New Zealand) would be prepared to consider the relinquishment of national rights and claims in Antarctica if such a step towards the establishment of a wider regime were generally agreed. This does not mean that New Zealand is in any way doubtful about the validity of its title to the Ross Dependency, or is unwilling to maintain its responsibilities for the administration of that area . . . Legal title has been supported by administrative acts to regulate whaling and other activities in the territory, and in more recent years by further exploratory and scientific

> expeditions . . . This vast and icy waste of 6 000 000 square miles, lying more in darkness than in light, has no permanent inhabitants. It can be explored and developed only with the resources of our advanced civilisation . . . It is only now being threatened by the tensions and rivalries that disturb the peace of the world elsewhere. (SPRI Archives 1959)

The views of Walter Nash were closely represented in this opening statement by the New Zealand representative. Nash's political priorities for the management of the Antarctic continent can be summarized: Any new international regime for the Antarctic would have to permit access to all nations intent on pursuing peaceful and/or scientific activities. Any new international regime created by the interested parties would have to have the blessing of the United Nations. Claimant states should consider relinquishing their claims if it would assist the development of a central authority for the management of the continent. Nash had, however, already indicated that he would be prepared to retain New Zealand's claim if the United States pressed a claim or the proposals for an international regime failed. Nash informed the New Zealand Parliament in 1960: "What we, as a Government desired on behalf of the people of New Zealand, was the abandonment in its entirety of claims by nations having or asserting sovereignty over any area, thus making it a completely international area without any possibility whatsoever of war" (cited in Alley 1984:914–15). The Antarctic had to be a demilitarized place with a prohibition on all nuclear activities.

The view of the New Zealand delegation was widely interpreted by other Commonwealth delegations as the least dogmatic in terms of territorial claims, freedom to conduct scientific research and access to the continent (Dodds 1996b). In their report on the Washington Conference, the New Zealand delegation noted that the Conference had been relatively free of "political polemic" because the superpowers had managed to co-operate effectively. In contrast, "Several of the participants at the Conference were openly unhappy about the whole idea of a Treaty, and the possibility of complete failure of the negotiations was never far removed . . . [although] the Chileans, Argentines and or French never pressed their objectives to the point of breakdown" (cited in Dodds 1996b). Through a system of formal committees and informal meetings, the points of objection such as the status of national claims or the possible incorporation of the high seas were gradually resolved. In their characterization of the participating parties, the New Zealand delegation identified three major groups: the ultra-conservative group who were anxious to protect their territorial claims and resource rights (Argentina, Chile, France and Australia), the big powers who were intent on securing a working international regime which acknowledged their substantial interests (the UK, the USSR, the USA) and the moderates who were the least dogmatic over territorial claims and legal rights (Belgium, South Africa, New Zealand and Japan).

Articles IV and V of the Antarctic Treaty, relating to national claims and nuclear testing respectively, were welcomed by a New Zealand government anxious that this geographically proximate region should be completely demilitarized. In conjunction with Argentina, Chile and South Africa, New Zealand pressurized the USA to accept that even peaceful nuclear testing was unacceptable either on the continent or the high seas. Given that Commonwealth records for that period frequently refer to the "steam roller" tactics by the superpowers in their quest for an agreement over freedom for scientific research and the legal suspension of territorial claims, the eventual agreement of the USA to the complete ban on nuclear testing was a considerable diplomatic triumph, even though these tactics closely affected the South American nations and France. The question of mineral resources within the region was not addressed, as other southern hemispheric nations such as Australia and Argentina objected to such a move, fearing that it might compromise their future sovereign economic rights.

New Zealand as a regional power: security in the Southern and Pacific Oceans

From the mid-1950s onwards, New Zealand's involvement in the Antarctic was shaped by the perception that it should be active in a region that was only a few thousands miles to the south. During the Cold War, sense of distance and even political responsibility had been altered by the development of long-range bombers, intercontinental missiles and the growth of global airlines. Far from making geography or territory irrelevant, these elements of time–space compression were heightening concern over local and regional issues. As Roderic Alley noted: "The uncomfortable possibility of superpower military rivalry at its doorstep, the jeopardising of what was an increasing commitment of resources for scientific endeavour by claimant/non-claimant squabbling, or even the possible weakening of New Zealand's claim in the Ross Dependency were all worrisome" (1984: 915).

The evolution of the New Zealand Antarctic programme has been characterized by a number of elements:

(a) New Zealand has relied upon the United States for logistical support in the Ross Dependency.
(b) New Zealand has been an active participant in the ATS and in the case of the negotiations over minerals in the 1970s has been active in promoting change within the ATS.
(c) New Zealand's role in regional defence organizations such as ANZUS and the ongoing controversies over the South Pacific have been an

instrumental element in shaping her concerns that the Antarctic remains a nuclear-free zone of peace (see Dalby 1993).

The role of the United States in facilitating the New Zealand Antarctic programme was highly valued by the New Zealand Antarctic community. The New Zealand Scott base is located within two miles of the American base (McMurdo) on Ross Island and for most of the last 30 years, the Americans have provided logistical support for the New Zealand Antarctic programme whilst her scientists worked on projects such as the Cenozoic Investigation of the Ross Sea (CIROS). In return the New Zealand government provided staging facilities in South Island. But as Auburn has noted: "France and the USSR have supplied their Antarctic bases without staging posts. It was estimated ten years ago [ie 1962] that the cost of a major all-year Antarctic base would be one to three hundred million dollars. This is clearly beyond New Zealand's resources, and for this reason Scott base has been built two miles from McMurdo" (1972:59).

Whilst some commentators have rightly highlighted the fact that New Zealand has enjoyed considerable savings (eg Alley 1984, Barber and Selby 1984), they have neglected to acknowledge the fact that New Zealand was a Cold War ally of the United States: "Meanwhile New Zealand gains many advantages from the US presence in the Dependency. It keeps the USSR out" (Auburn 1972:75, Hill 1983). New Zealand's dependence on US icebreakers was continuous between 1954 and 1983. Every year, US ships on their way to the Antarctic would call into New Zealand ports and collect New Zealand personnel heading for Scott base. The icebreakers would then spend up to two weeks clearing a channel through the ice, enabling HMNZS *Endeavour* to supply the Scott base (Auburn 1972:59–61). The president of the New Zealand Antarctic Society observed that: "Surely after over 25 years continuous operations in Antarctica, this country should be able to service its own outposts on the continent without having to rely on other nations to get materials and personnel to where they are required". There were also fears that the New Zealand government was failing to support the objectives of the Antarctic Treaty by refusing to purchase an ice-strengthened ship (Alley 1984). As Auburn noted 25 years ago, "Should the US decide to close down McMurdo, New Zealand would be practically incapable of operating Scott base" (Auburn 1972:60).

Fears of dependence rallied support for the creation of a regional defence organization at the start of the Cold War which was to tie together Australia, New Zealand and the USA against the apparent threat of communism or even a resurgent Japan. Contrary to popular belief, the ANZUS agreement did not provide a guarantee of US military support in the event of war in the South Pacific region (Dalby 1993). In the 1980s, however, the New Zealand government began to question the wisdom of defence arrangements such as ANZUS and for the first time refused to countenance nuclear weapons on their soil or in "their" waters. In the *Annual Report* of the Ministry of External Relations and Trade in 1988–9, it was reaffirmed that a prime goal of New Zealand foreign policy was to: "keep New Zealand nuclear free, and to promote adherence to existing instruments in support of a non-nuclear South Pacific and Antarctica" (New Zealand, Ministry of External Affairs 1989).

Whilst not claiming a direct correlation between the growing unease over Reagan's nuclear policies and the threat of disruption to the logistical support for the New Zealand Antarctic programme in 1983, American commentators were considerably irritated by New Zealand's lack of commitment to the nuclear option (Harris 1991). In 1985, John Dorrance (a US Consul General in Sydney) wrote that alliance solidarity was crucial given the necessity for nuclear co-operation and deterrence: "there are those who fear the dangers of engagement. They are led to an understandable concern about the horrors of war to argue that a policy of non-involvement or isolationism is the way to save our countries from a nuclear holocaust. Unfortunately, we are all now physically within the reach of nuclear weapons and their secondary effects . . . the first requirement must be to maintain our collective strength to deter aggression" (Dorrance 1985:217).

The decision of the New Zealand government to prevent US nuclear armed ships from visiting ports such as Christchurch and Wellington was unpopular in the United States strategic and foreign policy communities:

> With words New Zealand assures its partners that it remains committed to ANZUS; by deeds New Zealand has effectively curtailed its defence co-operation and operational role in ANZUS. The US has reluctantly adjusted its military relationship to the new reality, while New Zealand remains a valued friend its behaviour is no longer that of a close ally. A defence agreement has little meaning without defence co-operation – without some equity in sharing both the burdens and the rewards, New Zealand can't have it both ways. (Dorrance 1985:229)

According to Simon Dalby there were few operational necessities for such visits, but on a political level these "visits" were seen as confirming New Zealand's support for a US-dominated "free world". Whether or not Dorrance is considered as representative, there is no doubt that the Reagan administration was concerned with maintaining military alliances at the height of the Second Cold War (Dalby 1993).

The severance of the US–New Zealand logistical link occurred because the New Zealand government was increasingly concerned that the South Pacific and the Southern Ocean were rendered insecure by the nuclear policies of the United States. The Lange government suspended the ANZUS agreement in 1987 when the USA refused to confirm or deny the presence of nuclear weapons on board a vessel. A more independent and peace-oriented foreign policy was a shift away from the hitherto dominant agendas initially of Britain and latterly the United States (Dalby 1993).

Lange argued that New Zealand should be looking to its neighbours in the South Pacific rather than responding to the dictates of the North Atlantic Treaty Organization. Indeed the cover of the 1985 Green Paper on defence had a map of New Zealand constructed in such a way as to emphasize the proximity of the country to the South Pacific and the Antarctic rather than to the northern hemispheric nations such as the USA (Dalby 1993). The subsequent creation of a South Pacific Nuclear Free Zone in 1987 complements the nuclear free zone of the Antarctic. Calls for a nuclear free Pacific were first crystallized at a conference in 1975 attended by the microstates and islands of the Pacific. The decision to suspend ANZUS also coincided with the political creation of the Nuclear Free and Independent Pacific Movement which campaigned on themes of maldevelopment, independence and environmental security. Lange's decision to increase defence linkages with the Australians in the late 1980s was one sign that New Zealand intended to promote a regional view of the world sensitive to its needs and not those of its northern allies.

These connections were consolidated by the Bolger government in the mid-1990s. At the Pacific Forum summit in August 1996 at the Marshall Islands, the New Zealand delegation proposed that a Southern Hemispheric Nuclear Free Zone be created by linking the existing nuclear free zones of the South Atlantic, the Indian Ocean, the South Pacific and the Antarctic. This proposal appeared particularly poignant given that at this meeting the Marshall Island government noted a proposal to turn several contaminated atolls (67 nuclear explosions had been carried out by the US between 1946 and 1958) into repositories for nuclear waste.

New Zealand's resource interests in Antarctica and the Southern Ocean

New Zealand has been an influential member of the ATS in terms of addressing the living and non-living resource question. Her role has frequently been interpreted as existing between the hard-line territorial concerns of claimant states such as Argentina and Australia and the non-claimant states such as the United States and Japan. At the 1970 ATCP meeting in Tokyo, the New Zealand delegation raised the issue of resource management in the Antarctic. Other southern hemispheric nations such as Australia and Argentina were hostile to this development, fearing that their sovereign economic rights might be compromised in the process of devising a common approach to resource regulation. In contrast, New Zealand's delegation attempted to push for a decision on whether exploitation would be permitted in the future or banned permanently regardless of whether American and/or other research vessels detected traces of hydrocarbons in the Ross Sea.

The general discussion over resources in the 1970s coincided with fears of rising populations, increased potential for conflict and diminishing resources. Against this backdrop, the New Zealand government proposed at the Oslo ATCP meeting in 1975, having first raised the issue at the 1970 ATCP meeting, that Antarctica be declared a "world park" in which the mining of resources was prohibited. That approach failed, and a subsequent review of foreign policy acknowledged that:

> Some . . . favour a total moratorium on any form of commercial exploration or exploitation. Others see carefully controlled exploitation as both desirable and inevitable. A solution probably lies somewhere between these two opposed, and strongly held views. In a world of diminishing food and energy resources a moratorium is unlikely to represent a permanent or long term answer; nor is it likely to be an attractive solution to the international community at large. On the other hand, the idea of unilateral exploitation, with the greatest advantage accruing to the most technologically capable, is equally unacceptable, for it runs headlong into what have been hitherto irreconcilable difficulties over sovereignty and access, and raises profound environmental impact problems. (*New Zealand Foreign Affairs Review* 1977:100)

At a special meeting of SCAR in Paris in 1976, the Antarctic community agreed to a mining moratorium until the creation of a new minerals regime. New Zealand had suggested that certain areas of the Antarctic region be designated as off-limits for mineral exploitation – including areas of special scientific and resource interest. She also recommended the creation of a resource regulatory committee which would be responsible for maintaining a system of regulation (Mitchell 1980). Geopolitically speaking, these early debates were important in transforming the Antarctic from a place exclusively dedicated to science to one invested with considerable resource potential.

Undoubtedly, New Zealand was one of the strongest supporters of the parallel discussions on a convention for marine living resources which had taken place in the late 1970s. The signing of CCAMLR in Canberra in May 1980 was supported by the government in Wellington. Along with Norway and the United Kingdom, New Zealand called for this regime of resource management to be strengthened by demanding rigorous rights for inspection of fishing catches and the observation of resource exploitation targets, rather than voluntary compliance (Joyner 1992:227). The reluctance of the ATS to impose tougher controls over resource exploitation could have been driven by fears that strict resource management might imperil the ATS itself, given the vague jurisdictional definitions of CCAMLR. As with other claimant states, New Zealand interpreted the Convention as protecting her rights both above and below 60 degrees South (Auckland District Law Society 1979). This was confirmed by the 1977 Territorial and Exclusive Zone Act which made provision for the declaration of an exclusive economic zone for the Ross Dependency. The New Zealand government supported CCAMLR not least because, as before, it meant that the resources of the Southern Ocean would be subject to some form of regulation, especially given the large fishing fleets of the Soviet Union, South Korea, Poland and Japan (Beck 1986:231).

The mineral negotiations of the early 1980s coincided with renewed concerns that the resources in the Antarctic might be commercially exploitable in the near future. New Zealand had been involved in some forms of offshore drilling in the Ross Sea region. The minister of the environment, speaking in 1983, confirmed that: "The area could be among the first to attract the attention of the international oil companies. At issue is the right of New Zealand to control the exploitation of oil and gas in that part of Antarctica which is claimed by us" (cited in Hill 1983:39). The construction of a minerals regime had been "kick-started" in 1981 due to growing pressures from the ATS membership to find a solution to the regulation of mineral exploitation. The parties agreed at Buenos Aires in 1981 that the goals and principles of a draft agreement on mineral regulation would have to be agreed upon in the face of outside pressures such as the UNCLOS negotiations (Beck 1986:249). The key article was Recommendation XII:

> Any agreement . . . should be acceptable and be without prejudice to those states which have previously asserted rights of or claims to territorial sovereignty in Antarctica as well as to those states which neither recognised such rights of or claims to territorial sovereignty in Antarctica nor, under the provisions of the Antarctic Treaty, assert such rights or claims.

Territorial sovereignty was the key issue as questions of resource management were elevated to the forefront of Antarctic business. Amongst the claimant states were considerable differences of opinion, ranging from New Zealand's willingness to negotiate a new management regime to Argentina's reluctance to discuss the issue of sovereign control. The Law of the Sea (LOS) negotiations had complicated matters still further as it questioned the issue of control of common heritage resources and revenue sharing (Suter 1991). The United States had already rejected the LOS proposals for control on the mining of minerals in the deep seabed region. There was also evidence that environmental movements were becoming increasingly interested in the future management of the Antarctic. As early as 1973, the World Conference on National Parks had declared that the Antarctic should be a World Park under United Nations control (Suter 1980).

New Zealand, as an energy dependent nation, was concerned with these discussions given that some New Zealanders considered, "Antarctica is valuable potentially as real estate . . . the world's last frozen asset, an unexploited continent with ocean approaches heavily laden with sea life" (Barber and Selby 1984:466). Yet, in his opening statement to the 1981 ATCP conference in Buenos Aires, Chris Beeby stressed that any minerals regime for the Antarctic would have to fulfil three major objectives: the protection of the Antarctic environment, accommodation between the claimant and non-claimant states, and assurances that any new regime would also have to ensure that there were benefits to the wider international community (ATCP 1981). These sentiments were consistent with an earlier working paper prepared by the New Zealand delegation in 1979 for the Washington ATCM which stressed that any international resource regime would have to have a commitment to: environmental protection, an agreed regulatory body, anti-pollution guidelines, sharing arrangements for revenue derived from mining and a process of dispute settlement. Moreover, the New Zealand government, through a Ministry of Foreign Affairs discussion paper, was also aware that the future stability of the ATS was at stake: "It is the view of the New Zealand authorities that, as long as the Consultative Parties have not been able to conclude a minerals regime and so fill the gap in the Treaty about resources, the latter are poorly placed to resist outside pressure . . . It is for this reason, above all, that New Zealand is taking a lead and pressing for early progress on a minerals regime" (Hill 1983:73).

New Zealand's involvement in the CRAMRA proceedings was exemplified by the role played by Christopher Beeby, the head of the New Zealand delegation, who had been designated the spokesperson for the claimant states and subsequently chair of the minerals negotiations. The 1983 ATCM special meeting in Wellington was crucial to those proceedings as members of the ATS struggled to reconcile territorial rights with a minerals resource regime (Chaturvedi 1994:15). An early draft of a possible minerals agreement had been accepted for discussion even though, as Beeby confirmed, "it is quite clear that this negotiation will not settle the sovereignty question" (cited in Alley 1984:926). Indeed in the opening address to the 1983 ATCP special meeting in Canberra, Beeby returned to this important theme:

> In recent years, the Consultative Parties have come to grips with the resource issues that are left unresolved by the Antarctic Treaty . . . That development has seen an increase in the interests of the international community at large in our work. Interest has been accompanied by questioning, by criticism and by misunderstanding . . . We have not done enough to explain to others the nature of the Antarctic Treaty System in general and the purpose of our examination of the mineral issue in particular . . . The negotiations on minerals are not based on the knowledge, nor even the conjecture, that there are great riches to be harvested in Antarctica . . . Against the possibility, however, that exploitation might one day be found to be both feasible and environmentally acceptable, the Consultative Parties have made a judgement that it is important to reach agreement in advance . . . Our purpose is to fill the gap in the Treaty which, if left, could bring back rivalries the Treaty was designed to overcome. (ACTP 1983)

The New Zealand delegation continued to play a delicate role in shaping these negotiations over a minerals regime through the production of revised versions of the original Beeby draft (Orrego Vicuna 1988). In contrast to Argentina and Chile, their delegation emphasized the importance of negotiating a legal framework which would protect the Antarctic environment but not prejudice the interests of the wider international community. Indeed the expanding membership of the ATS in the 1980s was seen as a source of strength by New Zealand, in so far as important states such as Brazil, India and China were participating in discussions over the minerals issue. It was hoped that the growing Second and Third World membership of the ATS would deflect some of the criticisms of the Malaysian-led common heritage campaign in the United Nations. As the head of the New Zealand delegation, B. Thompson, noted at the 1985 ATCP meeting in Brussels: "We in New Zealand are well aware of the increasing interest in Antarctica . . . and need therefore for the Antarctic Treaty partners to tell the rest of the world more about themselves and the excellent work they are doing" (ATCP 1985).

The notion of cultivating better relations with the wider international community was considered beneficial in the face of mounting opposition to the Beeby proposals for a minerals regime. The idea of a regulatory committee overseeing mineral activity attracted particular criticism since it would have been composed of only eight member states. As Peter Beck noted: "Each committee would consist of the sponsoring state for the exploration application, any claimant to that area, three other claimants chosen by the first claimant, plus additional non-claimants chosen by the Chairman of the Commission" (Beck 1986:257). The apparent favouring of the claimant states not only caused some disquiet amongst the non-claimant ATCPs but also added fuel to the Malaysian-led campaign in the United Nations throughout the 1980s. The minerals question became an important plank in the campaign of the General Assembly to dismantle the ATS and to declare the Antarctic a common heritage of mankind (Triggs 1986).

The formal adoption of CRAMRA in June 1988 was the product of months of negotiations over the legitimate rights of ATCPs balanced with measures to protect the polar environment. The New Zealand delegations to the ATCP meetings had consistently stressed the latter element of that combination. At the Rio de Janeiro meeting of October 1987, the delegation had noted that: "We would like to highlight the priorities for New Zealand at this Consultative Meeting. The first two relate to the issue of environmental protection . . . the management of waste disposal in Antarctica . . . We also want to see substantial progress in the area of environmental impact assessment" (ATCP 1987). The pressure on the negotiating parties within the ATS had increased due to the UN-led debates over common heritage, condemnation of South Africa, the accession of new members such as Greece and the Koreas, the failure of CCAMLR to agree new conservation measures and the growing involvement of environmental movements hostile to CRAMRA. The Lange government had concluded in 1987 that CRAMRA had to be ratified in order to maintain the stability of the ATS (*New Zealand Foreign Affairs Review* 1987). New Zealand's reasoning throughout the CRAMRA years was consistent in the sense that it recognized that the environmental protection of the Antarctic had to be the number-one priority rather than narrow sovereign interests over mineral exploitation and the authority of the ATS itself.

At the formal conclusion of the CRAMRA negotiations, Chris Beeby predicted that the signing of CRAMRA would be, "an historic occasion which I believe will go down in Antarctic history as the most important political development regarding the regulation of the Antarctica since the Antarctic Treaty itself was adopted in 1959" (cited in Kyriak 1996:38). By June 1988, stumbling blocks to the ratification process emerged. The Australian delegation, critical of provisions for revenue sharing and the anti-subsidy clauses rather than the environmental provisions of CRAMRA, did not offer formal congratulations to the negotiating teams in contrast to the other ATCP parties. The strength of the criticism from environmental bodies such as the Antarctic and Southern Ocean Coalition (ASOC) and Greenpeace remained substantial. Greenpeace, for example, announced in 1988 that: "This is the first step to commercial exploitation of Antarctica and is one of the most shameful moves the New Zealand government could make" (cited in Kyriak 1996: 38). Cath Wallace of ASOC was a key figure in the anti-CRAMRA movement in New Zealand and later argued that: "by providing property rights the Convention will facilitate mining and the introduction of rivalry over strategic resources in the zone which will introduce new and dangerous stresses to the Treaty and area. This would endanger not only the environment but could lead to conflict and destabilise the Treaty [as well as] the zone's demilitarised status" (cited in Kyriak 1996: 71–2). The New Zealand branch of ASOC was scathing over the Lange government's involvement in CRAMRA, especially given earlier proposals for a Antarctic World Park in 1975, whilst Malaysia and other Third World states issued statements condemning the CRAMRA negotiations and final convention. On 26 November 1988, New Zealand, in conjunction with Norway and South Africa, signed the convention. By 22 March 1989, Australia and France were the only two remaining claimant states not to have signed the CRAMRA document.

The decision by Australia and France to boycott the minerals convention in May 1989, however, effectively ended CRAMRA. Michael Kyriak, who had access to New Zealand government files for the period between 1988 and 1991, has recently noted that the key months leading up to the October ATCM in Paris were dominated by the issue of formulating a response to Australian and French objections to CRAMRA (Kyriak 1996). In May 1989, the New Zealand government had decided that CRAMRA was the preferential option in terms of dealing with the minerals issue, as there was considerable ambiguity in both the French and Australian statements. At this stage it appeared that the French did not object to CRAMRA to the same extent as the Australian government of Bob Hawke which had announced its intention of seeking further measures for the environmental protection of Antarctica. The New Zealand government had notified Australia in June 1989 that they would resist a public campaign against CRAMRA and later agreed that the final decision on ratification of CRAMRA would be delayed until the Paris meeting. After several lengthy meetings of interested government departments, a White Paper on the Antarctic Environment was issued in August 1989 which announced government strategy. It re-established that the number-one priority for Wellington was regional security and: "that it [Antarctica] remains peaceful and nuclear free. The Antarctic Treaty is a superb disarmament model for the world" (cited in Kyriak 1996:4). More importantly, it proposed that security against military threats would now have to be cemented through measures designed to promote environmental security in the region.

Whilst ASOC's Cath Wallace condemned the White Paper as a cynical public relations exercise in August 1989, the geopolitical significance of this paper lay in the recognition that environmental security for New Zealand would only be secured if an integrated and binding protection regime was negotiated by the ATS. Key provisions would include: the regulation of tourism, marine pollution controls, heritage protection, the creation of protected areas and the development of environmental impact assessments. It concluded that, "CRAMRA cannot be viewed as the full answer" (cited in Kyriak 1996:11). In the months leading up to the Paris ATCP conference, the New Zealand government proposed the creation of a secretariat to improve the running of the ATS and a Standing Working Group on the Antarctic Environment. The latter received some support from Chile, Norway, Brazil and South Africa at preliminary meetings for the ATCP meeting.

At the Paris meeting of the ATCPs in October 1989, Chris Beeby confirmed that:

> It is vital that we do make progress here in Paris so that we discharge our responsibilities to Antarctica, to each other and to our publics . . . We need to look to a more comprehensive and integrated regime of environmental protection for Antarctica . . . New Zealand's support for such a regime has been expressed in a White Paper on the Antarctic environment . . . That Paper states a clear commitment by the New Zealand government to the elaboration of an integrated set of measures . . . to secure effective and comprehensive protection for the Antarctic environment. (ATCP 1989)

New Zealand subsequently submitted a working paper on the environmental protection of the Antarctic to the Paris meeting at the same time as the Australian and French delegations submitted their proposal for "Comprehensive measures for the protection of the Antarctic environment and its dependent and associated ecosystems". The pressure on the ATCPs was considerable in the aftermath of the CRAMRA negotiations. Argentina and Norway were not prepared to compromise on the importance of sovereign rights in the face of growing international interest in the region, whilst others such as the United States were considering devising new measures for environmental protection (Chaturvedi 1996:134). Subjected to a constant barrage of criticism from ASOC, the New Zealand government was forced to react positively in the post-CRAMRA months of 1988–9 (see *ASOC Antarctic News* November 1989).

The campaign for a permanent ban on mining coupled with a comprehensive convention on environmental protection was discussed at a special ATCM meeting held in Vina del Mar during November–December 1990. The previous meeting in Paris had failed to bridge the profound gaps between states over the future of CRAMRA and/or a new environmental agreement. New Zealand's position was significantly different from its strongest Antarctic allies – the United Kingdom and the United States. In contrast to the geopolitical reasoning of the northern hemispheric nations, the New Zealand government did not concentrate on sovereign rights and territorial claims. It was recognized within the New Zealand Ministry of Foreign Affairs that the growing internationalization of Antarctica, demonstrated by the involvement of a coalition of 200 environmental NGOs in Vina del Mar, meant that the ATCPs would have to negotiate a new comprehensive environmental regime which encapsulated a permanent ban on mining. Whilst there were differences between ATCPs in Vina del Mar, the delegations managed to secure agreement on two major issues: first, a protocol format was established in part to facilitate the addition of amendments and, second, an environmental advisory body would have to be created in order to supervise implementation.

The key negotiating role of Trolle Andersen of Norway during these negotiations was substantial given objections from the UK, USA, Chile, Argentina and the Soviet Union to proposals for a permanent ban on mining. During the negotiations, the Palmer government was under considerable political pressure to push for a permanent ban given that the National Party and ASOC had called for a complete ban. Ironically, after months of negotiations in the aftermath of the Vina del Mar conference, New Zealand remained the only party pressing for a complete ban on mining. The so-called Group of 4 (Australia, France, Belgium and Italy), which had initially called for such a development, later accepted a temporary prohibition and other states such as Japan also agreed to a review-based moratorium. The negotiating strength of the USA and the UK meant that the New Zealand proposals designed to protect any possible overturning of the mining ban were considerably weakened in the final development of a Protocol on Environmental Protection.

New Zealand strongly supported the Protocol on Environmental Protection during the Madrid ATCP meeting in June 1991. The period between the Vina del Mar meeting and the Madrid conference had seen significant changes in the attitudes of the negotiating parties. New Zealand's delegations were successful in modifying the text of the Protocol, especially in terms of the moratorium on mining and the need to negotiate a new legal regime. The Protocol designates Antarctica as a "natural reserve devoted to peace and science" and binds its present and future signatories to total protection of the Antarctic environment. It also categorically prohibits any activity relating to mineral resources. Christopher Beeby, who had chaired the mineral negotiations between 1982 and 1988, noted that the search for a resolution on the mineral and environmental questions was a product of a belief that:

> The single most important stimulus towards agreed rules was the common judgement that their absence might one day lead to a revival, in a very acute and unmanageable form, of the dispute about sovereignty put aside by Article IV and that this could, in turn, severely undermine the Treaty and even lead to its collapse. That judgement – that rules are needed, that the gap in the Treaty should be filled – still represents common ground. (Beeby 1991:18–19)

This sentiment reflected a long-standing concern of New Zealand that the Antarctic Treaty represented the first stage in the development and evolution of the international political and environmental management of the polar continent and surrounding oceans. The subsequent creation of the International Centre for Antarctic Information and Research (ICAIR) in Christchurch in 1991 could be interpreted as one part of New Zealand's planned contribution to this general objective. The Centre is committed to the collection of scientific and environmental information on the Antarctic and Southern Ocean. Under the management of New Zealand, Italy and the United States, the ICAIR is intended to contribute to the development of a greater number of environmental assessments in the Antarctic region.

In the last four or five years, the New Zealand government has pursued a twofold strategy with respect to the polar region: first, the ratification of the Protocol was considered a priority and was achieved through domestic legislation in December 1994. Throughout this period, New Zealand had pressed for the Transitional Environmental Working Group to develop management structures which would enable the implementation of the Protocol and argued that issues such as the regulation of tourism and the question of a secretariat would have to be addressed in the near future. New Zealand proposed at the 1995 Seoul ATCP meeting that information on tourism should be shared amongst interested parties and that the effective management of tourism would require the ATCPs to foster close co-operation with tour operators. The geopolitical reasoning underlying these priorities was based on the assumption that the ATS would build upon the successful negotiation of the Protocol and tackle other issues that might either compromise the environmental protection of the continent or invite further criticism of the ATS. This is a highly pertinent issue given the repeated attacks on the ATCPs by ASOC and other environmental groups concerning the slow ratification of the Protocol by Treaty parties.

A recent development which deserves comment was the decision of the Bolger government to create a new agency in 1995 to manage the Antarctic programme: New Zealand Antarctic Institute (Antarctica New Zealand, ANZ). The latter was assigned four key roles: the implementation of the New Zealand Antarctic programme, the environmental stewardship of the Ross Dependency, the provision and distribution of information about policy and the raising of public awareness of Antarctica amongst the New Zealand public. Old discourses about territorial sovereignty and resource control appeared to have been replaced by environmental stewardship/security and commercial exploitation of tourism originating from Christchurch. With this administrative re-organization, it was hoped that inter-governmental co-ordination would be improved, given different responsibilities for the Ross Dependency Research Committee than those for the New Zealand Antarctic programme, the Department of Environment and the Ministry of Foreign Affairs and Trade (MFAT). In July 1996, ANZ was transferred from the administrative structure of the MFAT to become a stand-alone crown agency responsible for implementing government policy. The striking aspects of recent publications from ANZ is the recognition that scientific activity is no longer the sole defining activity of the Antarctic and consequently will have to co-exist with other issues such as commercial tourism and the development of Christchurch as a major gateway port for the region. Moreover, environmental stewardship for the Ross Dependency is to be combined with an appreciation that: "the conservation of the intrinsic values of Antarctica and the Southern Ocean for the benefit of the world community and for present and future generations of New Zealanders" is a priority for the New Zealand government (cited in Prior 1997: 21). New Zealand's future involvement in the Antarctic will be increasingly measured in terms different from those developed in the late 1950s.

Summary

This chapter has considered the historical and political significance of New Zealand's role in the Antarctic and the Southern Ocean. The historical context for this region is embedded within the changing nature of Anglo-New Zealand relations and reactions to the activities of the United States and Norway. The first half of the present century was characterized by political apathy on the part of New Zealand governments which tended to favour minimalist foreign and economic policies. In the post-war period, interest in the Antarctic region expanded largely as a consequence of political pressure from other Commonwealth states. In the last 30 years, successive governments have been staunch defenders of the Antarctic

Treaty System, and their delegations have stressed the importance of developing resource agreements and environmental protection measures for the region. Whilst New Zealand has been a passionate defender of the ATS, attitudes towards further international co-operation and the involvement of non-state actors have differed considerably from those of Argentina and Chile. Successive governments of New Zealand have encouraged greater participation in the belief that the continued operation of the ATS is ultimately a higher priority than the promotion of territorial rights and/or resource exploitation. However, the opening up of the Ross Sea fishery in February 1997 (approved under CCAMLR) will place new pressures on the New Zealand government to go ahead and declare an EEZ, which will include delimiting the seabed and continental shelf, around the Ross Dependency in order not only to regulate the sustainable fisheries management but also to ensure access for New Zealand's commercial fishing interests (Prior 1997).

Within New Zealand, the recent discussion of the country acting as a "gateway" nation to the Antarctic and the Southern Ocean is a reflection of changing geopolitical and strategic perceptions. In the light of the recognition that the southern polar policy makers of the northern hemisphere have either diverted funding to other geographical regions (the Arctic in Russia's case) or have been subjected to financial constraints (the USA), there have been renewed calls for: "the need, or desirability, of closer coordination and interaction of the five Southern Hemisphere Gateway countries to Antarctica: New Zealand, Australia, Chile, Argentina and South Africa" (Prior 1997:4). In conjunction with its membership of the Valdivia Group, New Zealand has been active in promoting scientific and diplomatic contact with neighbouring southern hemispheric states on areas of mutual concern such as environmental management, resource exploitation, polar tourism and Antarctic science. The protection of the Antarctic environment in the 1990s is, therefore, just one element in a wider southern hemispheric portfolio.

8

South Africa and the Antarctic

The long history of South African participation in the Antarctic and the southern seas is a neglected if important dimension of the Republic's history. The urban fabric of Cape Town and Durban bear witness to their logistical support for expeditions and as centres for a South African whaling industry:

> South Africa's involvement with the Antarctic continent and the sub-Antarctic Prince Edward Islands can be said to have commenced with the first explorations of the southern seas . . . Later Cape Town was a major port for many southern sealing and whaling expeditions and to a number of the "heroic age" expeditions to the Antarctic continent. Only after the Second World War did South Africa become formally involved when it established a station on the Prince Edward Islands in the Indian Ocean. (Cooper and Headland 1991:77)

In 1948, South Africa annexed the sub-Antarctic Prince Edward Islands and this had implications for South African perceptions of the Southern Ocean and the Antarctic continent. South Africa was an original signatory of the 1959 Antarctic Treaty, and her participation in the ATS has spanned decades of international condemnation over apartheid and the illegal occupation of Namibia. The ATS was one of the few international forums in which South Africa could participate without fear of expulsion, in spite of attempts by outside parties in the 1980s to remove her. South African involvement in the Antarctic also touched upon other regional developments in the Southern Ocean, especially the proposals to create a South Atlantic Treaty Organization (SATO) as well as strategic alliances in the 1970s and 1980s.

This chapter initially considers the role of South Africa within British imperial polar policy. In contrast with Australia and New Zealand, the Union was a minor player in terms of Antarctic exploration and

Commonwealth polar co-operation. This was largely a consequence of limited economic resources and strained relations in the aftermath of the Boer War. Debates ensued whether South Africa would press a formal claim to the Antarctic in the 1930s were indicative of this condition. Whilst there were opportunities to follow Australia and New Zealand, the South African authorities were poorly organized and politically ambivalent about a possible claim. The annexation of the Prince Edward Islands in 1948 was significant because it fundamentally altered South African views on the Antarctic and Indian Ocean. After 1945, strategic assessments of the Southern Ocean and the Antarctic were increasingly predicated on the possible intentions of the Soviet Union in the region rather than resource and/or scientific evaluations. South African participation in the 1959 Washington Conference is reviewed together with the reasons for their participation and their evaluation of the Antarctic Treaty. South African interests in the Southern Ocean and the Antarctic are considered with respect to the resource issues of the 1960s and 1970s in conjunction with strategic fears of Soviet maritime expansionism. As with Argentina, the South African military contributed to the shaping of these perceptions, which often ran counter to the assessments of the scientific community. This account also considers the significance of the UN debates on the "Question of Antarctica" in alliance with the mineral resource debates. South African membership of the ATS became a political component of the critiques of Third World states such as Malaysia and Zimbabwe. Finally, in the post-apartheid era, the South African ratification of the Protocol on Environmental Protection is considered along with current concerns for greater environmental and resource co-operation in the South Atlantic.

Imperial Polar Policy and the South African Antarctic sector

The reasons why South Africa never pressed a territorial claim to the Antarctic continent reside within the complicated histories of Commonwealth relations and the international politics of the Antarctic (Laverde 1988, Dodds 1995, 1996b). South Africa's role in the imperial administration of the Antarctic in the 1920s and 1930s is crucial in this respect. To put it simply, the British authorities consistently argued that the portions of the Antarctic "closest" to Australia and New Zealand were considered of greater strategic and resource importance than Coats Land or Queen Maud Land. South Africa was, therefore, not invited to the earliest meetings between the Colonial Office and the representatives of Australia and New Zealand in 1921. In contrast to the Australians, South African participation in the Antarctic and the Southern Ocean during this period had been modest. Unlike abortive South African attempts to launch an Antarctic expedition, Sir Douglas Mawson had led several Commonwealth expeditions after the 1912 Scott expedition. Moreover, the situation between Britain and South Africa was probably not assisted by the bitter memories of the 1901–03 Anglo-South African war.

South African political interest in the Antarctic during the 1920s was often sporadic, poorly organized and lacking in funding. The main department for organizing South African Antarctic policy was the Department of Commerce and Industries rather than the Department of External Affairs. This reflected official concerns over the regulation of its commercial activities in the Southern Ocean. The South African-based Kerguelen Sealing and Whaling Company and the Union Whaling Company were engaged in a struggle to match the catching potential of Norway and Japan. The former company had negotiated in 1926 a 10-year lease from the British government to exploit the whaling, sealing and mineral resources on and around the Prince Edward Islands. The lease was later relinquished in 1934 at the request of the company (Cooper and Headland 1991:78). In other areas of official interest such as meteorology and climatic change in the Southern Ocean, there were signs of modest activity. In 1921, the South African Association for the Advancement of Science organized a committee in order to prepare for an Antarctic expedition. Four years later, after much delay and procrastination, General Jan Smuts issued an appeal for a scientific programme dedicated to Antarctic and Southern Oceanic observation. He noted that, "If the Union and Australia could agree each to maintain an Antarctic station opposite, or to the south-west of their respective territories, and the work of these Antarctic stations could be coordinated, the results might be of the utmost value" (Smuts 1925:1). This appeal coincided with negotiations with the British over the creation of a weather station on Gough Island. Whilst journals such as the *South African Nation* (March 1925) argued that the South Africans should participate in this type of research and expand on to other southern islands such as Tristan da Cunha or Gough, the government later declined to create a meteorological programme due to the onset of economic depression in the inter-war period.

The Imperial Conferences of 1926 and 1930 established the cornerstones of British and imperial polar policy. South Africa sent representatives to the Polar Committee established in London and designed to co-ordinate expeditions and advise on policy formulation. The primary objective of British polar policy was to consolidate titles to territory in the Falkland Islands Dependencies and the Ross Dependency. Other geographical areas of the Antarctic continent were also discussed at the first meeting of the Committee on British Policy in the Antarctic: namely, Coats Land and Enderby Land. The former territory, named after the Scottish Coats family, was geographically proximate to South Africa. However, British officials had concluded that this region (defined as 20 to 16.5 degrees West) "should be brought under the same administration as the Falkland Islands

Dependencies". The British strategy for maintaining and expanding upon existing claims over the Antarctic rested upon encouraging other possible Commonwealth claimants to explore and annex territories outside British jurisdiction. The irony of the situation was that South Africa, a recipient of British imperial policy, was not willing or able to press for a South African Antarctic claim.

The reasons for this apparent indifference were never outlined explicitly in government papers belonging to the Department of Commerce and Industries and/or External Affairs. However, it is possible to identify three possible reasons for this apparent apathy. First, successive administrations were preoccupied with nurturing the Dominion in the face of economic dislocation and the onset of global depression. Second, it was becoming apparent that the regulation of the whaling trade based on territorial control was unravelling, as the development of pelagic whaling meant that ships could operate for longer periods at sea without the support of land bases. British and imperial policies of resource regulation in the Southern Ocean were increasingly under threat from Norwegian and Japanese whalers which operated outside the jurisdiction of the imperial state. British attempts to persuade South African whaling companies to play a stronger role either in the regulation of resource exploitation or the possible creation of new claims drew sceptical conclusions from government departments. Officials attached to the Division of Fisheries, part of the Department of Commerce and Industries, were also aware of the New Zealand government's inability to regulate whaling in the Ross Dependency. Third, South African suspicion of British motivations in general remained in spite of the fact that the Anglo-South African War had ended some 25 years earlier.

The key moment in South African polar participation came with the news that the British had agreed to give up their claims to Bouvet Island in favour of the Norwegian government in the late 1920s. In November 1928, the British government had urged Norway not to claim any territory already under the formal jurisdiction of Britain and/or the Commonwealth. In his review of a high-level meeting between the Norwegian minister in London and Sir Robert Lindsey of the Foreign Office, Lord Cushendun of the Foreign Office reminded his audience that:

> The discussion of the whole Antarctic question was addressed . . . there were two aspects; the first was political and territorial. Considerations of geography alone, with the British Dominions of South Africa, Australia and New Zealand projecting down towards the Antarctic continent, gave the British Empire a very special interest in the political rights of those areas, particularly in view of the growing dispositions of the Dominions to interest themselves in the disposal of Antarctic territories. (cited in Dodds 1996a)

The British government, according to Lord Cushendun, was anxious to restrict further annexation of the polar continent to the jurisdictions of the

Dominions rather than outside parties. The common-sense appeals to geographical proximity, later to be used by South American rivals in the FID, were perceived to be counter to the territorial ambitions of northern hemispheric states such as Norway and the United States. However, the British were aware that the Bouvet question, in alliance with concerns over international whaling, meant that it was unlikely that the Norwegian government would accept such appeals.

The significance of the ongoing Anglo-Norwegian negotiations over Bouvet Island were appreciated by South African officials. The British were later to use concessions over the sovereignty of Bouvet Island in return for the withdrawal of Norwegian claims to Canadian Arctic islands such as Axel Heiberg. The Union government was aware, therefore, that their interests in the Antarctic and the Southern Ocean were only one element in a wider diplomatic and political game spanning the two polar regions (Dodds 1996a:29–30). At the 1930 Imperial Conference, the Polar Committee concluded that in the face of rising Norwegian interest in territorial claim making, there should be a concerted effort to divert that interest either towards the so-called "South African Antarctic sector" or to a distant south western sector. There were no formal South African objections to this proposal. The Union government, between 1925 and 1930, had neither dispatched a major Antarctic expedition to the "South African sector" nor raised any objections to the British policy of dealing with Norwegian insistence over Bouvet Island (Laverde 1988). The only major contribution in the early 1930s was the dispatch of the whaling ship *Tafelberg* to the waters of Dronning Maud Land as part of the South African participation in the 1932–3 International Polar Year (Cooper and Headland 1991). A year after the *Tafelberg* voyages, the South African political scene was reconstituted as Smuts and Hertzog decided to create the Union Party in an attempt to reconcile the bitter wounds of the Boer War and the problems of economic dislocation (Barber and Barratt 1990). It should be remembered that although Britain remained the dominant external influence on South African foreign policy, suspicion of British motives regarding the polar regions lingered. South Africa's ambivalence over Antarctic claims in the 1920s were later to be regretted, as officials in the 1930s worried that the Union had frozen herself out of active participation in Antarctic affairs.

Claiming a South African Antarctic sector?

In January 1935, a memorandum prepared by the Department of External Affairs was dispatched to other government departments for comments and reactions. The memorandum was entitled "The Union and the Antarctic" and acknowledged that:

> The attention of the South African public has from time to time been drawn to matters concerning the Antarctic region by notices and articles in the Press. So there appeared in *The Star* of the 3rd February last, an article suggesting that steps be taken by the Union government to gain control over the part of the Antarctic "geographically known as the South African sector" . . . The sector in question is officially described as the Sector between Enderby Land and the Falkland Islands Dependencies, i.e. roughly between 45 degrees E longitude and 20 degrees W longitude. (cited in Dodds 1996a)

The memorandum later conceded that any claim to this sector would depend upon the activities and political responses of the United States, Norway and others such as Japan. However, the importance of this document lies in the recognition of growing awareness within the government and the popular media that South Africa should be pressing a claim to the Antarctic continent. This renewal of interest in the Antarctic coincided with a number of events that bear mentioning. First, British and Commonwealth territorial hegemony had been challenged (Beck 1986, Peterson 1988, Klotz 1990). The Americans had encouraged the aerial endeavours of Richard Byrd and Lincoln Ellsworth. More significantly, it raised the possibility that the United States might be contemplating a territorial claim to the Antarctic in the near future (Klotz 1990). There was some speculation that this might have profound implications for New Zealand and the Ross Dependency too. Second, Argentina and Chile had reminded various bodies such as the International Postal Bureau of their territorial rights in the Falkland Islands Dependencies (FID). Finally, there were renewed concerns that the activities of Japanese and Norwegian whalers might contribute to over-exploitation of living resources of the Southern Ocean.

In the light of these changing circumstances, it was perhaps not surprising that the South African government began to review its position on either a possible Antarctic claim and/or South African policies towards the region. The first proposal for a South African polar expedition came from a British explorer, Captain Mills Joyce, in June 1930. In a letter to the South African High Commission in London, Joyce warned officials of the dangers of Norwegian polar involvement: "the Norwegians are displaying marked activity" (cited in Dodds 1996a:30). He expressed the hope that an expedition could be launched from South Africa that would conclude with "the South African flag be[ing] unfurled over those lands to the south of her continent" (ibid.). There was little official enthusiasm for this project in 1930. In London, officials at the Dominions Office were reluctant to embrace this proposal because it was feared that Anglo-Norwegian relations could be affected. The Polar Committee, however, was prepared to support a scientific expedition to the region which would have supplemented the long-standing endeavours of the voyages of *Discovery II* which had been a past visitor to the port of Cape Town (Ommanney 1938).

Remarkably, given its private origins, a Joyce-like proposal returned to the official Union agenda in 1935 when the Department of External Affairs produced a memorandum on the South African Antarctic sector. Three major reasons for annexation were offered within the document: "Opportunities for scientific research and meteorological observations . . . Possibilities of revenue from the whaling and sealing industries . . . The land is stated to be rich in minerals, guano and phosphates". It was proposed that:

> It would not be, it is suggested, too audacious to advocate the desirability of establishing at least one base from which South African enterprise could operate and expand its activities in case future developments justify such a course on a purely economic basis. This objective could be achieved by the annexation of some island, or possibly part of Coats Land extending into the South African sector . . . [such an action would] obviate the possibility of a loss of opportunity which may later prove to be a matter for regret. (South African Archives, Department of External Affairs 1935)

The "threats" facing this South African Antarctic sector were increasingly being defined in resource issues rather than in territorial terms. There were repeated concerns that Japanese whaling was compromising the living resource levels in the southern seas closest to the South African mainland. Fears of Japanese over-exploitation were undoubtedly related to wider concerns about the Japanese state in the aftermath of the invasion of Manchuria in 1931 and the bombing of cities such as Shanghai in 1931–2. The Fisheries Survey, in October 1935, warned the External Relations Committee in Pretoria that:

> It is advisable to inform the Norwegian Minister through the High Commissioner for the Union in London that Japanese whaling activities in the Antarctic are causing the Union government some concern and that, whilst not wishing to disturb any tentative agreement which the Norwegian government may have with the British government, South Africa is directly interested in the South African sector and the Union government does not feel inclined to allow the Japanese to operate in that area whilst withholding from establishing, say, a meteorological or marine biological research station on the Antarctic continent. (South African Archives, Department of Treasury 1935)

The identification of resource-led challenges to the South African sector were not effective in persuading the Union government to dispatch a major Antarctic expedition. The South African whaling firm Irvin and Johnson continued to operate in Antarctic waters, but Union officials did not instruct the company to press a territorial claim on behalf of South Africa. In part, this was based on the realization that the Norwegian whalers were already active in exploring new lands and charting new harbours.

On the eve of the Second World War, therefore, South African interest in the Antarctic and the Southern Ocean was undoubtedly modest in comparison to other nations such as Australia, Britain and Argentina. The efforts of the Geological Society of South Africa and prominent academics such as A.L. Du Toit to raise financial support for an Antarctic expedition came to nothing (Berman 1968). In contrast, the British government had supported the 1934–7 British Graham Land Expedition, the Americans had sponsored the expeditions of Admiral Byrd and the Argentines had undertaken the annual relief of their meteorological station on Laurie Island. During the war period, the international politics of the Antarctic were fundamentally altered by the decision of two South American states to press claims to the Antarctic peninsula and the British decision to launch Operation Tabarin (Beck 1986, Dodds 1996b). These changes led to a fundamental policy re-evaluation in the immediate post-war period and prompted the annexation of the Prince Edward Islands in 1948 by the South African government.

South Africa, the Prince Edward Islands and the South Indian Ocean

The changing international politics of the Antarctic was the subject of intense concern for the three southern hemispheric Commonwealth nations between 1946 and 1948. In a letter from the secretary of state for Dominion Affairs in November 1946, representatives from Australia, New Zealand and South Africa were warned of the significance of events such as the recent South American claims:

> Recent important developments in the Antarctic have led us to consider our previous policy in relation to general British Commonwealth territorial claims in the area . . . Commonwealth and New Zealand governments would be advised to take early steps to strengthen their claims to Australian Antarctic Territory and to the Ross Dependency respectively, and that this could only be effectively achieved by the establishment of at least one permanent base in each sector. (Australian Archives, Department of External Affairs 1946)

Whilst this letter referred directly to the existing claims pressed by Australia and New Zealand, the South African government, and in particular the Department of External Affairs, did initiate a review process of Union interests in the Antarctic and the Southern Ocean. Initially, the Department of External Affairs in December 1946 suggested to the British government that South Africa would be willing to press a formal claim to the polar continent if it was judged useful in the light of South American competition over the Antarctic peninsula and the uncertainties of the US government. Whilst the British government did not respond positively to this proposal, it did recommend that the Prince Edward Islands, located in the southern part of the Indian Ocean, be formally transferred to the South African government at the earliest possible moment.

The agreement to transfer ownership of the Prince Edward Islands signified the concern of the British government that the Islands needed to be formally incorporated into either the administration of the Falkland Islands Dependencies (FID) or another Commonwealth state. Relative geographical proximity was considered to favour the South African government as the Prince Edward Islands were 1200 miles away from Durban and over 4000 miles away from the FID. The British in alliance with their Commonwealth partners were anxious that a series of islands scattered around the Southern Ocean should be secured, in terms of ownership and occupation, because of Cold War-induced fears over the intentions of other nations, particularly the Soviet Union. The eventual occupation of the Prince Edward Islands in 1948 by the South African government charted a new course in terms of their interests in the Antarctic and Southern Ocean. The strategic, legal, meteorological and resource implications were evident as a Department of External Affairs memorandum concluded in October 1947:

> In the opinion of the South African Chief of General Staff, modern long range weapons, the fact that accurate weather forecasting is of the greatest importance in warfare, and the present uncertain position in the Indian Ocean makes these islands of the greatest strategic importance to the Union . . . In the long term, however, the question arises that if South Africa does not lay claim to the islands now and the United Kingdom allows such a claim as it has to lapse, there is the possibility that they may be claimed by some other country that may not be friendly disposed towards the Union. For the proper defence of the Union, therefore, the islands must remain at least in friendly hands. It is felt that the only way to ensure this is for South African sovereignty to be established as soon as possible. (South African Archives, Department of External Affairs, 1947)

This statement of intention reflected the international vision of President Jan Smuts, a strong advocate of Commonwealth co-operation in the polar regions as well as a passionate defender of Western security interests in the southern African region. The transfer of the Prince Edward Islands to the Union government was, therefore, significant in cementing Smuts's belief that the British government recognized the importance of the country for Commonwealth collective security (Barber and Barratt 1990).

Whilst the actual details of the sovereignty transfer and the secret operation (Snoektown) that formally annexed the Prince Edward Islands (PEI) in January 1948 are reasonably well known, the annexation coincided with developments in the international politics of the Antarctic such as the Australian annexation of Heard Island which led the South African newspaper *Cape Times* to contend that: "With the occupation of Heard and the Prince Edward Islands the western powers have secured a solid belt of islands which will be used as bases for air communications in the southern seas" (3 January 1948). This line of geopolitical reasoning was further reiterated within the South African Parliament as government ministers legitimated the Prince Edward Islands annexation (see Marsh 1948, *South African Government Gazette Extraordinary* 1948, Cooper and Headland

1991, Dodds 1996a). The Minister of Finance, F. Sturrock noted that: "We were concerned to establish a meteorological base there . . . There was also a danger of these islands coming into the hands of unfriendly powers, and that they would be a serious menace to South Africa in these days of rocket bombs and other projectiles which could be fired from there. For these and other safety reasons we felt that the annexation was justifiable" (cited in Dodds 1996a:35). The appeal to strategic dramas and to Cold War depictions of threatening "others" was ultimately successful when the South African Parliament supported the annexation expenditure and the creation of a permanent meteorological base.

The American proposals to internationalize the Antarctic, released in August 1948, were not dispatched to the South African Embassy in Washington. This was controversial because the Union government, since the annexation of the PEI, considered herself to be an active participant in the region. The American decision to exclude the South Africans from consideration was based on the belief that only existing claimant states should be involved in the negotiations over the future of the polar continent. At that stage this limited consultation process was designed to restrict the Soviet Union from this process rather than South Africa *per se*. In a response to the Department of State proposal, the South African Embassy in Washington reminded the former of the implications of this omission:

> The Union has not hitherto advanced any claim to Antarctic territory most of which it has regarded as *terra nullius*. However, South Africa is the only major power in the Southern Hemisphere which is apparently to have no say in the future control and administration of the Antarctic continent. Control, when it is exercised, will doubtless prove to be of vital concern to the Union . . . Finally in the field of long range and long term strategy, the control of Antarctica will always be a matter of major concern to South Africa. (cited in Bush 1982:198)

The response of the Truman government was not unsympathetic in tone but nonetheless reiterated its belief that the consultation process should only include claimant states for the present time. Moreover, it was widely appreciated that South Africa would have to be kept informed of any political developments through the liaison networks of the old Commonwealth states.

The Cold War and the growing international isolation of South Africa in the 1950s shaped the political culture of South Africa's Antarctic interests. Whilst it would be fair to conclude that the polar continent was not the highest strategic priority, it was not considered in political and geographical isolation. The Department of External Affairs under the leadership of Eric Louw played a key role in ensuring South African participation in the Antarctic and the Southern Ocean in the midst of four major developments: the maintenance of a research base and weather programme on Marion Island; participation in the 1955–8 Commonwealth Trans-Antarctic Expedition and a financial commitment of £18,000; participation in the research programmes of the International Geophysical Year; and finally the negotiations leading to the take-over of the Norwegian station in the Antarctic (La Grange 1991). As with their Australian counterparts, the South African governments of Malan and Strijdom were hostile to the various plans for internationalizing the Antarctic and concerned that the Soviet Union might establish research bases in the region.

Throughout the 1950s, the South African government faced ignominy and uncertainty as to whether a formal invitation to discussions concerning changes to the international political circumstances of Antarctica would be forthcoming. British officials deemed South Africa, as a non-claimant state, to be less important than Australia and New Zealand and when the South African representative De Villiers, at a Commonwealth meeting in September 1956, mentioned the possibility of South Africa being involved in any polar discussions, there was no official response from Britain or Australia (Dodds 1996c). New Zealand, however, did acknowledge that South Africa had a geographical interest in the region. Regarding South Africa as a pariah state, condemned over its policies of apartheid and the illegal occupation of Namibia, the Eisenhower administration was reluctant to include her in any emerging talks. Moreover, there was concern that if South Africa were included then the Soviet Union might insist upon participation. As a Commonwealth Relations Office memorandum reproduced for Commonwealth foreign ministries seemed to confirm in September 1957:

> It appears that the United States is not keen on having the South Africans participate in these conversations lest recognition of its interests might become known and serve as a pretext to the Russians to press for eventual participation. South Africa would very much like to attend and is pressing more and more for recognition of its interests in the Antarctic. (South African Archives, Department of External Affairs 1957)

In the run-up to the International Geophysical Year (1957–8), negotiations between the American administration and the Commonwealth states continued over possible changes to the international politics of the Antarctic. In a meeting with the Commonwealth Relations Office in July 1957, the South African representative raised the issue of possible Soviet involvement in the Antarctic. The minutes of the meeting record that South Africa was reluctant "to accept a situation which might open the way for the Soviet Union to establish bases in the Antarctic". Both the South African and Australian delegations were adamant that the Soviets should not be allowed into the Antarctic. The virulent anti-communism of these two states sat uneasily with the more pragmatic British response that the Soviet Union had established bases in the Antarctic and was unlikely to leave given their logistical investment and articulation of a possible claim on the basis of earlier exploration (Toma 1956). Moreover, the Commonwealth could not rely on American support because uncertainty existed regarding the latter's intentions to press a formal claim to the continent. South African fears over Soviet involvement could be considered misplaced given that the British were undecided whether to pursue a plan for a new international body involving only the claimant states or a wider agreement involving the superpowers and possibly South Africa.

In the period between the annexation of the PEI and the International Geophysical Year (IGY), the Dominion was an active participant in the Southern Ocean and took part in polar expeditions ranging from the 1949–52 Norwegian–British–Swedish expedition to the Commonwealth Trans-Antarctic Expedition. A scientific programme had been established and administered by the Department of Transport, the Weather Bureau and the Geological Survey of South Africa. The decision of the Soviet Union to remain in the Antarctic in the aftermath of the IGY was a crucial moment in the transformation of South African fortunes. From 1958 onwards, it was apparent that nations such as the Soviet Union would have to be included in discussions over the future of the polar continent. The decision to extend an invitation to the South African government for the preliminary discussions on the future management of the region in 1958 was a diplomatic triumph for the Verwoerd government, which had earlier been censored in the United Nations over apartheid and the illegal occupation of Namibia. Given South African concerns over the possible intentions of the Soviet Union in the Antarctic, the Department of External Affairs in Pretoria was swift to accept this invitation from the Eisenhower government (Cooper and Headland 1991, Dodds 1995).

South Africa and the Antarctic Treaty

In a statement to the South African Senate on 17 March 1959, Foreign Minister Eric Louw reported that South Africa's policy towards the Antarctic was influenced by the geographical fact that as a proximate nation it had a special relationship with the polar continent. He reminded colleagues that Queen Maud Land was only four hours away by plane from the South African mainland and that modern nuclear technology meant that missiles could be launched from the continent towards South Africa. This theme was reiterated again in Louw's opening address to the Washington Conference. His speech encapsulated the three major aims of the South African delegation at the conference: to avoid the involvement of the United Nations in the international politics of the Antarctic, to secure the demilitarization of Antarctica and to ensure that existing claims to the continent would not have to be formally recognized by South Africa. Moreover, the South African delegation was also anxious to place the Antarctic within a strategic drama derived from the Cold War. Louw told the 11 other delegations in October 1959 that:

> We therefore welcome the proposal of the US government, that a conference be held which could draw up a multinational Treaty embodying two fundamental principles; that Antarctica shall be used for peaceful purposes only, and secondly that there shall be freedom of, and international co-operation in, scientific investigation in Antarctica . . . By virtue of South Africa's geographical situation at the southern end of Africa, and thus the nearest part of that continent to Antarctica, it stands to reason that we in South Africa regard events and developments in the Antarctic continent – our southern neighbour – as of particular importance and of direct and special concern to us. . . . The strategic importance of Antarctica has become increasingly evident in recent years. Today South Africa is only a few hours flying time away from the Antarctic continent. The main, in fact the only sure sea route between the east and the west is around the Cape of Good Hope. Thus, not only for South Africa, but also for all peace loving nations, it is imperative that this sea route remains open to all nations. (SPRI Archives 1959)

The South African polar programme became established during the Washington Conference. The *New York Times* reported that Norway had loaned the South African government her base on Queen Maud Land and the use of Bouvet Island meteorological station (8 December 1959). As a consequence, the South Africans approved the creation of a South African National Antarctic Expedition, and in December 1959 the expedition left Cape Town and travelled to the former Norwegian base in Dronning Maud Land. On 8 January 1960, the first South African polar base was established and formally renamed SANAE.

South African ratification of the Antarctic Treaty coincided with her formal membership of the Scientific Committee on Antarctic Research in June 1960 (Cooper and Headland 1991). As an independent body unattached to the United Nations, South Africa's participation was based on her scientific contributions to polar research (especially during the IGY) rather than on the nature of her political system. In doing so, South African diplomats and scientists had unprecedented access to the international polar community and opportunities to forge diplomatic and research linkages. South African–Norwegian co-operation over the transfer of an Antarctic base is a case in point, even though Norway had condemned the Republic for apartheid in the United Nations in 1962. The *Cape Times* reported the Norwegian Ambassador to America, Paul Koht, commenting that "I can assure our South African friends that Norway takes great pleasure in being able to help the cause of science in this way" (*Cape Times* 16 October 1959). In general, therefore, South African agreement to the Washington Treaty precipitated membership of SCAR and access to the international polar community. The Council for Scientific and Industrial Research (CSIR) co-ordinated the South African Committee on Antarctic Research (SACAR) which represented South Africa in SCAR. The Department of Transport was invested with command of the Antarctic programme.

The South African government was swift to ratify the final Treaty document in June 1961. It resolved the issues that were of pressing concern such as declaring the Antarctic a demilitarized place and banning nuclear explosions and testing from the continent. South Africa had been a firm supporter of the Argentine delegation's proposal for a total ban on nuclear testing. Territorial claims had been suspended under this agreement and South African policy of the non-recognition of claims had not been compromised; by signing this treaty South Africa had not implicitly accepted the territorial claims of Australia or Argentina. Science was declared the principal activity for the Antarctic continent. As a southern hemispheric nation, the Antarctic Treaty had ensured that the polar lands were removed from the dramas of the Cold War and, moreover, possible Soviet intrigue. As the *Cape Times* concluded, "12 nations, including South Africa and the Soviet Union, yesterday signed an historic treaty to preserve the Antarctic for peaceful scientific research for all time, banning all military weapons from the 5 000 000 square miles of continent" (2 December 1959).

South Africa and security in the Southern Ocean

During the 1959 Washington Conference, the South African government had been in negotiations over the possible creation of a naval pact between either South American states or former imperial powers with interests in southern Africa. Whilst South African scientists were preoccupied with programmes of research in the Antarctic and the Southern Ocean, successive governments were anxious that South Africa should secure the shipping lanes around the Cape of Good Hope and the open spaces of the southern seas. The strategic assessment of the region was undoubtedly shaped by the Cold War and fears of Soviet maritime expansionism. South African Minister of External Affairs Eric Louw was instrumental in advocating a regional defence organization for the South Atlantic in the late 1950s. The impetus for subsequent diplomatic and military manoeuvres was precipitated by the northern hemispheric nations', such as Britain and the United States, reluctance to enlarge the remit of the North Atlantic Treaty Organization. Southern hemispheric nations such as South Africa and Argentina became increasingly frustrated that the defence of the Southern Ocean had been neglected and that unless this strategic vacuum was filled, there was a possibility that the Soviet Union could destabilize Western shipping.

The central relevance of these developments lies in the way that the changing dramas of the Cold War influenced South African representations of the South Atlantic and the Antarctic. South Africa's heightened role in the polar continent should not be divorced from wider regional concerns of Soviet maritime expansionism and the defence of the Southern Ocean. Whilst South African governments would have been reluctant to jeopardize their involvement in the ATS by contravening the provisions of the Antarctic Treaty governing demilitarization, there is evidence to suggest that the Antarctic was considered part of a wider strategic picture. In his opening address to the Second Consultative Party meeting in Buenos Aires in July 1962, Johan Maree outlined South Africa's perceptions of the Antarctic:

> We, South Africans, living on the southern tip of the continent of Africa, have long regarded Antarctica as an important factor in the great storms which bring sudden and dramatic changes in our weather . . . South Africa's primary interest in Antarctica has traditionally been in the weather, but in recent years we have been able to maintain a station and extend our studies to include associated transient phenomena connected with the inosphere, cosmic rays, geomagnetism and the aurora . . . There is also much scope for co-operation in the political field . . . Soviet and Japanese Antarctic vessels of other Treaty powers often make use of Cape Town's harbour and of the facilities available there for the calibration of instruments. (ATCP 1962 cited in SPRI Archives)

The South African delegation was anxious to promote scientific and political co-operation over matters such as logistical support, the administrative structure and telecommunications (Kingwill 1971). One reason for this interest could have been based on increasing fears of widespread isolation in other international forums such as the United Nations General Assembly, and therefore, a willingness to promote co-operation where possible such as in the ATS was imperative.

More broadly, a central theme in South African foreign policy in the apartheid era was to stress the importance of the Republic to the general defence of the 'West' (Sidaway and Simon 1993, Dodds 1995). South Africa holds a large amount of the world's strategic minerals deemed vital to Western industrial development. "South Africa holds very large regional, continental and global significance. It is strategically what the military vocabulary calls 'vital ground'. It is the key to the countries south of the Zambezi, it is an arsenal of industrial capacity for the free world and an ideological citadel difficult to invade by communist propaganda" (Gygli 1976:180). The Cape of Good Hope was a major point from which thousands of ships passed through on their way from the Middle East to Europe and North America. The introduction of the supertanker in the 1960s meant, for example, that passage through the Suez Canal was no longer an option. South Africa had also become a buffer against Sovict expansionism in Africa. South African governments, especially after Vorster, argued that Pretoria had to improve her diplomatic and military relations with other southern hemispheric nations such as Argentina and Chile.

Within the Antarctic Treaty System, South African delegations had a range of diplomatic and scientific contacts unparalleled in any other international organization. The issue of apartheid interfered neither with the running of the South African Antarctic programme nor with the participation within the ATS or SCAR. The priority for successive South African governments in the 1960s and 1970s was to maintain South African participation in the ATS so that the Republic could contribute to the administration of the Antarctic and maintain a strategic interest, which encapsulated environmental and resource security, in the region. The Department of External Affairs was also anxious to participate in the regulation of living resources in the Southern Ocean. The resource management of krill and whales had been substantial during this period, either through participation in the BIOMASS programme and/or aerial and sea-based surveying of Southern Right whales off the African continent (Miller 1991a, 1991b). South African delegations at the ATS became increasingly anxious over the possible exploitation of non-living resources after several ATCP meetings in the 1970s. As the leader of the South African delegation to the Washington ATCP meeting in September 1979 later noted: "I think there can be no doubt that unless the treaty nations reach agreement among themselves within the foreseeable future on these issues the possibility or likelihood of uncontrolled mineral activities will increase dramatically . . . a legal framework must be devised and accepted which would regulate any future mineral activity in Antarctica . . ." (ATCP 1979).

Whilst ATCPs were prepared to co-operate over Antarctic matters, South Africa could not rely upon international support outside the closed meetings of the ATS. One way South Africa tried to strengthen her international position was by means of a defence pact with Argentina (see Forrest 1981 for details on South African–Brazilian linkages). Both states were dominated in the 1960s and 1970s by authoritarian and anti-communist governments. In 1967, South Africa began a series of exchanges with the Argentine and Brazilian armed forces in the hope of exploring the possibilities of creating a naval pact. This followed an earlier programme called the SANARA (South African Navy and Armada Republica Argentina), a semi-official organization dedicated to the promotion of good relations between the South African and Argentine navies (Dodds 1995). In June 1969, the commander-in-chief of the Argentine navy, Pedro Gnavi, visited South Africa after Vice-Admiral H. Bierman of the South African navy had visited Buenos Aires in 1968. The purpose of these meetings was to consider the potentials for creating a South Atlantic Treaty Organization (SATO).

Proposals for a South Atlantic defence pact emerged on the South African political agenda throughout the 1970s. New fears were expressed that the southern seas around South Africa would be vulnerable to Soviet expansionism. Some events were deemed to be critical to these hardening perceptions: the oil crisis of 1973–4, the decolonization of Angola and Mozambique, the withdrawal of the British from Simonstown and internal uprisings following the Soweto killings (Hanlon 1986, Henshaw 1992). Admiral James Johnson of the South African navy argued in 1976 that "The communists are turning the area into a Soviet lake . . . On any one day you can see 30–35 Soviet ships pass by here and there's nothing we can do. We are all alone" (cited in Hurrell 1983:183). Military officers in Pretoria and Buenos Aires expressed the hope that a defence pact could be created in order to prevent Soviet maritime expansionism coupled with aggressive actions directed from the front-line black African states. South African political leaders were intent, therefore, on identifying dangers in the Southern Ocean and the African continent (Dodds 1994a). The articulation of a total national strategy (TNS) was one element of the Botha government's plan for resisting internal dissent and aggressive front-line states (Geldenhuys 1981). As the Commander General of the South African Defence Force, H. Biermann, over-stated in 1973:

> Communist penetration into the southern hemisphere and the threats that this portends have caused the southern hemisphere, and particularly the Indian Ocean, to emerge dramatically from a position of relative obscurity and to assume a conspicuous position in the East–West power struggle . . . In addition [to naval co-operation with South American navies] there is significant co-operation in respect of scientific expeditions to the Antarctic. This co-operation is important and efforts should be made to promote and to extend this within the framework of the defence of the southern hemisphere. (cited in Dodds 1995:65)

On the one hand, defence commanders such as Biermann recognized the importance of the ATS in allowing South Africa to foster diplomatic and logistical relationships; on the other hand, the Antarctic was considered part of a regional strategic picture. Ironically, of course, the South African government proved perfectly capable of developing not only a considerable arms arsenal but also of executing policies designed to destabilize southern Africa (cf. Moorcraft 1981, Hanlon 1986). Moreover, the South African government invited a number of governments to consider either the formation of a defence pact or defence co-operation. These other nations included Argentina, Israel, Paraguay, Uruguay and Chile (Hurrell 1983).

The much vaunted South Atlantic pact did not materialize in spite of a high-level conference in Buenos Aires in 1981. Since 1976, various attempts to construct such a SATO had failed for a number of reasons: interested parties such as Brazil refused to co-operate with South Africa, South African–US relations were in a parlous state because of the US support for a UN resolution banning arms sales to South Africa, and Argentina was becoming increasingly preoccupied with internal economic and political crises. South African involvement with the 1982 South Atlantic War was minor but not insignificant in spite of her neutrality. There was considerable speculation over three issues: the role of the South African naval base in Simonstown, arms sales to Argentina and the renewed possibility of a SATO. Officials in Pretoria were swift to scotch any rumours over South African involvement in the Falklands crisis. Pik Botha announced that: "The South African government maintains friendly relations with both the

United Kingdom and Argentina and sincerely hopes that the two governments will be able to resolve amicably the dispute over the Falkland Islands" (cited in Dodds 1995).

The significance of the Falklands campaign was to shape South African perceptions of the Southern Ocean and the South Atlantic in particular. Whilst it has been demonstrated that the activities of the ATS were unaffected by the campaign, the immediate impact can be traced to the decision of the South African navy to call a one-day conference in Pretoria in May 1982. The *Cape Times*, for example, reported that fears were expressed that the Soviet navy might take advantage of the conflict and impose a naval hegemony in the region. Whilst the extent of this possible Soviet role was always hazy (Argentina had cultivated a trading relationship with the USSR), it did illustrate in part persistent fears of the intent of the Soviet blue-water navy (*Cape Times* 12 May 1982). There was also much speculation inside the armed forces and within the media over the mineral potential of the Antarctic in the light of the forthcoming discussions over a minerals convention. R.K. Campbell of the Institute for Strategic Studies in Pretoria exaggerated the apparent issues at stake:

> The Falklands have shown that South Africa is, in effect, faced with a credible threat of invasion . . . What might the vastly more powerful Soviet navy achieve, with pre-planning and long term preparation . . . even more useful would be a small force of frigates . . . and to support South Africa's own, often forgotten, outposts and possessions such as Marion Island and the Antarctic base SANAE, South Africa's Antarctic ambitions are simply impossible without some sea-going naval vessels to support them. (1984:13 and 24)

In strategic terms, it appeared to many conservative commentators to confirm their fears that the Southern Ocean and even the Antarctic could become a source of conflict and/or tension. Whilst the Antarctic Treaty System was not undermined by the conflict, countries such as Britain expanded their involvement in the Antarctic in the face of renewed speculation over the possible exploitation of minerals in the Antarctic. The South African delegations to the ATCP meetings in Buenos Aires (1981), and later in Canberra (1983) were involved in proposing several draft clauses in the earliest stages of the CRAMRA negotiations. In particular, the South African delegation attempted to by-pass the problematic issue of the previously asserted rights to territorial sovereignty in Antarctica by some states against those which neither recognized these claims nor asserted their own rights or claims.

Furthermore, the Convention on Antarctic Marine Living Resources (CCAMLR) had entered into force in 1982 and purported to deal with the regulation of resources such as fish and krill (Miller 1991b, 1991c). The provisions of the CCAMLR encapsulated a number of sub-Antarctic islands such as Marion and meant that the South African government was affected by these provisions. The attempts to regulate the exploitation of living resources in the Antarctic during the 1980s remained problematic because of three major obstacles: first, South Africa in conjunction with other nations such as Argentina and France raised objections over the international regulation of national maritime space. The South African government was concerned that their sovereign rights around the Prince Edward Islands might be compromised. Second, CCAMLR relied on consensus decision-making, and the development of catch quotas depended upon the goodwill of a number of parties including the Soviet Union. Third, a lack of scientific data on the ecosystems of the Southern Ocean made the regulation and prediction of krill and fish increasingly dubious.

South Africa, the United Nations and the "Question of Antarctica"

International attention in December 1982 was primarily focused on the final stages of negotiation over UNCLOS III. It was no accident, therefore, that the legal status of another common space/property regime, the Antarctic Treaty System, became the centre of global attention. "Political interest in the Antarctic became widespread during the early 1980s among the Group of 77 in the United Nations, and since 1983 the issue of Antarctica has been included as an agenda item for the General Assembly" (Joyner 1990a:42). Malaysia and other nations argued that, in the light of pressing resource concerns and the needs of the developing world, the international regime of the Antarctic should be subject to investigation and renegotiation. The first announcement by the developing world on the future management of the Antarctic coincided with the meeting of the Non-Aligned Movement (NAM) in September 1983. The final document called for a more just and equitable framework for the exploitation and management of resources of the Antarctic. "The inherent logic of the new international economic order demands that the resources of Antarctica be available for the common benefit of mankind . . . It is only the United Nations that can play a role, representing as it does the interests of all member states" (UNGA 1983 cited in Dodds 1995:68).

The NAM declaration appeared well founded given the general acceptance of the common heritage concept for outer space and the high seas and the appeals of the 1980 Brandt Report which had advocated the idea of a New International Economic Order. The initial campaign by the developing world was characterized by demands that the Antarctic be declared a place of common heritage allied with demands for increased participation and democratization of the decision-making process of the ATS. According to the Malaysian representative at the United Nations the ATS, "has not kept pace with current international realities. The Antarctic Treaty System . . . is unacceptable because of its exclusivity, its unaccountability and its secrecy' (UNGA 1984). For many in the Group of 77, the ATS was understood as an exclusive and unaccountable international regime because the negotiation of the Antarctic Treaty in 1959, for instance, had preceded the creation of many African and Asian countries in the 1960s and 1970s from the fragments of European empires. The Office of the Secretary-General was persuaded in 1984 to commission a substantial report on the future status of the Antarctic. For the Group of 77 that was the starting point of an eight-year campaign to challenge the political status of the polar continent. Added urgency was given to these debates because, throughout these years, the ATS members were attempting to negotiate a minerals regime for the Antarctic.

In the initial stages of these General Assembly debates, the question of South African membership of the ATS was not considered to be of primary importance. However, in 1985, and largely as a consequence of African states such as Zaire and Kenya, the issue of apartheid and South African membership was raised in the General Assembly debates on the "Question of Antarctica". In United Nations resolution 140/156C, "The General Assembly . . . *Notes with regret* that the racist apartheid regime of South Africa, which has been suspended from participation in the General Assembly of the United Nations, is a Consultative Party to the Antarctic Treaty . . . *Urges* the Antarctic Treaty Parties to exclude the racist apartheid regime of South Africa from participation in the meetings of the Consultative Parties at the earliest possible date" (UNGA 1985, cited in Dodds 1995:69). The resolution was carried overwhelmingly by the General Assembly and those in favour included ATS Consultative Party members such as India and Brazil. Other members of the ATS such as Britain and Australia did not participate in the vote and argued that the UN had no authority to interfere with an existing legal regime.

The timing and significance of the "South African resolution" was based on several factors. The award of the Nobel Peace Prize in November 1984 to Archbishop Desmond Tutu, an outspoken critic of apartheid, undoubtedly drew world attention to the continued injustices of apartheid South Africa. The resolution, therefore, presented critics of the ATS with a further opportunity to contest the moral and political legitimacy of this organization. How could the morally repugnant apartheid regime of South Africa be a Consultative Party member of the ATS? Calls for South Africa's expulsion caused considerable embarrassment within the ATS community, with states such as Australia forced to support South African membership after having unreservedly condemned apartheid on other occasions. Richard Woolcott, the Australian ambassador to the UN, explained in 1984 that:

> The Australian government has unequivocally condemned and deplored the institutionalised racism, which the system of apartheid in South Africa involves. But the Antarctic Treaty is a functioning international instrument, which cannot be replaced or altered except by its contracting parties. In this connection, I

> should say, with reference to South African membership of the Treaty, this is really an issue for the Parties of the Treaty themselves to determine, as is the case in any other Treaty system. (cited in Dodds 1995:70)

The ATS membership would be forced to account for inconsistencies over South African membership, given that the preamble of the 1959 Antarctic Treaty agrees to further the "purposes and principles embodied in the Charter of the United Nations" (cited in Klotz 1990:195). This could be interpreted to imply that South Africa's membership was morally wrong because of her violation of human rights conventions via apartheid and because of international condemnation of the illegal occupation of Namibia (Sidaway and Simon 1993).

There is no question, therefore, that the issue of South African membership of the ATS provided a useful diplomatic "stick" with which to beat the ATS Consultative Parties. There was, for example, never any formal UN-based discussion of South African contributions to Antarctic science or its role in the ATS over the preceding 25 years. From the ATS's conservative perspective, expelling South Africa would establish the principle that outside parties could interfere with the running and membership profile of this legal and political regime (Beck 1986, 1994c). Whereas for the Group of 77, the South African issue could be mobilized in support of their critiques of the ATS both inside the General Assembly and outside in other international forums. At the NAM Conference in September 1986, the keynote address by Robert Mugabe of Zimbabwe commented that: "The Heads of State or government noted with regret that the racist apartheid regime of South Africa is a Consultative Party to the Antarctic Treaty, and in the light of UNGA Resolution 140/156C urged the Consultative Parties to exclude that regime from participation in their meeting forthwith" (cited in Dodds 1995:70). This was followed by a renewed onslaught in the General Assembly in November 1986. The Ghanaian delegate, J.V. Gbeho, noted that he was "surprised that those who professed to be allies against racism and apartheid had suddenly found the South African regime an indispensable partner in the Antarctic Treaty. The Treaty should not even be remotely seen as co-existing with racism" (*Southern African Record*, Number 46 1987:74). Other African states such as Zimbabwe, Rwanda and Nigeria later supported UN resolution A/C 1/41/L88 condemning South African membership of the ATS (Beck 1994c).

The consistent response of the ATS community between 1984 and 1992 was to reject these annual General Assembly resolutions calling for the ejection of South Africa. The South African government through its ambassador to the United Nations, Errol de Montille, continued to reject the legality of these resolutions. South African newspapers such as the *Star* (Johannesburg) adopted a defiant stance and argued that the Botha government could ignore the criticisms of other African states and Malaysia (see, for example, the issue of 24 February 1985). Whilst it was reasonably clear that South Africa could rely upon the support of the original 12 Consultative Parties, other members such as India and Brazil campaigned for South Africa's removal from the ATS. At the same time, ongoing controversies over the proposed minerals convention (CRAMRA) meant that South African membership was thoroughly immersed in these wider issues of resource exploitation and management. Some anti-apartheid observers such as South African geologist Maarten de Wit, suggested that: "it is possible further economic pressures could be brought to bear upon Pretoria to desist from the present racial policies by the threat to terminate the purchase of such minerals from South Africa and obtain supplies from Antarctica instead" (De Wit 1985:55). However, such a position seemed to run contrary to many sceptical commentaries about the mining potential of the Antarctic and the environmental, technological and political costs attached to such activities (see, for example, Cook 1990).

The South African government did not further their cause in the United Nations by attracting unwanted publicity over their plans for an airstrip on Marion Island. The proposal to construct an airstrip in 1986–7 was rapidly condemned by international and national environmental groups and governments. There was much speculation over the motivation for the construction of an airstrip close to the South African meteorological station and scientific base. The South African government argued that the airstrip would enable the evacuation of injured personnel, increase control of territorial waters and improve the possibilities of getting supplies to the base in an emergency. For some observers such as Greenpeace the airstrip proposal was indicative of the military contemplating either nuclear testing in the region or planning to improve communication facilities in case the Antarctic became a site of controversy over the mining issue. Other commentators such as South African scientists who worked on Marion Island pointed to the existence of a rocket testing facility in Cape Province and suggested that the military wanted to upgrade the facilities on Marion Island in order to enable the recovery of rockets fired from the South African mainland (Dodds 1995). It has been alleged, for example, that Israeli personnel travelled to Marion in 1984–5 to assist with the initial feasibility studies for an airstrip. Leaving aside the reasons for the proposed runway, there is no doubt that, had such an airstrip been constructed, massive ecological damage would have resulted (Anonymous 1987).

It must be remembered that the South African government had failed to inform either the international scientific community or the ATS membership about the proposal. The Prince Edward Islands are effectively part of the Antarctic Treaty system because they are incorporated into the provisions of the CCAMLR. Whilst there may be no legal obligation for the South African government to inform the ATS of its intentions, other ATS members would have expected to have been informed of their plans. In the event, the Department of Environmental Affairs later decided to drop this proposal after widespread criticism. In the formal government review of the Marion Island landing facility, the minister of Environmental Affairs conceded that, "I have decided that although a need does exist for a landing strip on Marion Island, to provide such a facility is not desirable because of the impact it would have on that fragile environment" (G. Kotze 12 May 1987, cited in Heymann et al 1987). Rather than a simple landing strip, the Botha government proposed that a 1400-metre runway suitable for C130 Hercules be laid and that landing facilities on the surrounding beaches should be constructed (Heymann et al 1987).

The significance of the Marion airstrip controversy should not be underestimated. The allegations about nuclear testing and/or rocket testing facilities raised concerns over the dominant South African representations of the Southern Ocean. Whilst many South African scientists would have rejected such proposals outright, the South African Defence Force and the Department of Defence were an important component of the former apartheid state. The South African government had created a considerable controversy at the height of the mineral regime negotiations. After six years of debate, the ATCPs had finally agreed a convention on the regulation of mining and mineral resources (CRAMRA) and declared it open for succession in November 1988. At that point, in spite of opposition from green movements, South Africa had been prepared to sign this convention because CRAMRA had made some provision for environmental protection and the careful regulation of mining. South African newspapers such as the *Star* (Johannesburg) had declared their support for the convention. "South Africa has a long standing interest in the fate of the nearest neighbouring continent . . . The exploitation of the Antarctic minerals is surely unrealistic and the environmentalists' dreams of a great snowy nature reserve are also unrealistic . . . South African interests are more than scientific; the continent has large deposits of coal, iron and gold, which, if exploited, could hurt the economy' (7 May 1988). The prospects for the ratification of CRAMRA eventually evaporated in May 1989 when Australia and France announced that they would not sign the convention and instead pressed for a comprehensive environmental protection agreement.

To be ratified, CRAMRA had to be approved *inter alia* by all the claimant states (including Australia and France). The consensus within the ATS over non-living resources had been shattered as other states such as South Africa and Britain argued for the retention of CRAMRA. Shortly after the Australian decision in May 1989, South African newspapers and environmental groups such as the Dolphin Action and Protection Group began campaigning against CRAMRA. By the time of the special meeting of the ATCP in Vina del Mar in November and December 1990, it was clear that the ATS was split over the issue with some states such as Norway and Britain pro-CRAMRA and others such as Sweden and New Zealand demanding a permanent ban on mining. At the time of the Vina del Mar discussions, UN resolution 45/78B of 6 December 1990 "[appealed] *once again* to the Antarctic Treaty Consultative Parties to take urgent measures to exclude the apartheid minority regime from participation in their meetings at the earliest possible date until such time that the abhorrent system and practices of apartheid domination are totally eliminated in South Africa" (UNGA 1990, cited in Dodds 1995:73). The resolution was passed by 107 votes (including ATCPs; India and Brazil) in favour and none against; the other members of the ATS announced their non-participation in the General Assembly vote.

The pressure of the developing world was imposed on the ATS in two overlapping areas in the early 1990s. In the first place, South African participation in the Antarctic Treaty System attracted further resolutions in favour of immediate expulsion. Second, the Group of 77 campaigned for the Antarctic to be declared a "common heritage of mankind". These calls were heightened as a result of the ATCP failure either to ratify CRAMRA or to develop a new resource convention in 1988–90. The ATS, however, in 1990 did agree to a draft protocol for a fixed moratorium on mining. It was decided that the protocol would depend upon the Madrid meeting of the ATCP in June 1991. The details of these discussions have been considered in other chapters of this book. South Africa did not play a major role in those negotiations. It was not until the Bonn meeting in October 1991 that a more comprehensive Protocol on Environmental Protection (Protocol) was created and later signed by South Africa. The South African minister of environmental affairs, Louis Pienaar, who signed the Protocol on behalf the Republic, was later reported as saying that: "The signing of this agreement in Spain last week will be of great benefit for the protection of the Antarctic environment and its associated ecosystems" (the *Star*, 15 October 1991).

The South Africans considered the Protocol an important milestone in the international history of the Antarctic (Mills 1994, Dodds 1995). The Protocol established a fixed moratorium on mining in the region for the next 50 years, and fears over the possible economic and environmental effects of the exploitation of minerals in the Antarctic have been frozen – for the present. The signing and ratification of the Protocol not only removed some of the political tension over the exploitation of the Antarctic and debates over common heritage but also coincided with the freeing of Nelson Mandela which meant that South Africa's pariah status had begun to wane. The last UN resolution to call for the expulsion of South Africa from the ATS was passed in December 1992 (Beck 1990a). However, the political salience of these calls had already begun to evaporate in the wake of the Protocol and internal changes in South Africa. The Protocol had declared the Antarctic a "nature reserve devoted to peace and science" and dealt with resource exploitation. The election of Nelson Mandela as president in 1994 meant that the Group of 77 could no longer use South African ATS participation as a political lever for debates over the political and legal standing of that international regime.

After apartheid: South Africa in the Antarctic

The election of Nelson Mandela in 1994 marked the formal ending of the politics of apartheid. In conjunction with the ending of the Cold War, it is generally agreed that an era of international isolation and condemnation has been replaced by the expectation that South Africa will fully rejoin the international community, regaining access to a variety of political organizations, aid, international capital markets, social and cultural events. Since June 1994, South Africa has rejoined the Commonwealth, resumed its seat in the United Nations, joined the Group of 77, the Southern African Development Community, the Organization of African Unity, initiated negotiations with the European Union and signed major pieces of legislation such as the Framework Convention on Climatic Change (Davies 1995, Rowlands 1996, Vale 1996). Diplomatic relations since 1994 have been either restored or initiated with 78 other nations and the number of overseas representatives has increased.

Whilst the geographies of apartheid appear more resistant to change, the creation of a Government of National Unity (GNU) has undoubtedly been preoccupied with tackling these social and spatial inequalities. The ambitious Reconstruction and Development Programme is an important element in the post-apartheid era which will have substantial political and financial implications for South Africa (Lemon 1995). In the circumstances, it is perhaps surprising that South Africa's commitment to the Antarctic has not been unduly affected in the short term. One of the recent achievements of the Department of Environmental Affairs was to secure agreement from the African National Congress (ANC) in 1992 for the continued funding of the Antarctic and Islands programme after forthcoming democratic elections. In the same year, the ANC released a working paper on strategic priorities which declared that "As good environmental citizens, we will strongly support the comprehensive protection of Antarctica. We urge the development of instruments which will enable the continent to become a 'Nature Reserve – Land of Science'" (cited in Mills 1994:235).

In the last few years the prospects for South African participation in the Antarctic have remained fair in terms of funding and political support. The most substantial investment by the government has been the new R13 million SANAE IV Antarctic base to replace the disintegrating SANAE III in Queen Maud Land (see Map 8). The base was finally completed in the summer season of 1995–6. In combination with existing funding levels, South Africa will continue to pursue a range of scientific programmes including studies of the Antarctic atmosphere, the monitoring of the Queen Maud Land ice cap, geological research in the western Dronning Maud Land, oceanographic research such as participation in the World Ocean Circulation Experiment and meteorological research in Marion Island (*Environmental Affairs Annual Reports* 1992–3, 1993–4). Minister of Environmental Affairs Wynand Breytenbach noted that these studies would demonstrate that "Apart from the long term value to continue with the oceanographic, biological, earth and physical sciences research, South Africa's involvement in the Antarctic is of strategic importance." The minister did not elaborate on the strategic importance of the Antarctic. As with official pronouncements, it is widely assumed that South Africa's proximity to the Antarctic means that it has a natural interest in resource exploitation, territorial waters, climatic change and security in the Southern Ocean. Whilst this argument is perfectly understandable, it needs to be carefully considered rather than assumed.

An interesting development in the 1990s has been the articulation of new arguments about South Africa and security (Dodds 1995). Traditional arguments based on fears of the Soviet Union and ballistic missiles have been replaced by concerns over polar environmental security. In an interview with Dirk van Schalkwyk, a senior official attached to the Antarctic and Islands section of the Department of Environmental Affairs, he confirmed that South Africa's strategic interests in the region were currently concerned with changing ozone levels over the Antarctic (with implications for South African weather and levels of skin cancer), with global warming and the exploitation of living resources such as krill or whales. South African concerns for resource exploitation and possible climatic change are likely to remain the most powerful reasons for a continued commitment to research in the Antarctic and the Southern Ocean. Since 1979, South Africa has declared a 200-mile nautical fishing zone around the Prince Edward Islands in the face of rising concern of over-exploitation and international fishing competition (*Republic of South Africa Gazette* 1979). It also has to be remembered that if UNCLOS III were ever ratified by all the international community, the Prince Edward Islands could be invested with their own territorial sea, exclusive economic zone and continental shelf. The resource implications alone would mean that no South African government in the next 50 years is likely to reduce funding significantly for the Antarctic programme or for research on Marion Island. It might even extend existing legislation such as the 1988 Sea Fisheries Act and the 1989 Environment Conservation Act to protect the marine resources of the Islands.

In the immediate aftermath of the ratification of the Protocol on Environmental Protection in July 1995, the South African government faced considerable challenges in balancing its Antarctic interests with wider regional concerns such as SAZOPC, the Indian Ocean Basin Initiative and the Valdivia Group (Mills 1996, Venter 1996). The current attempts to

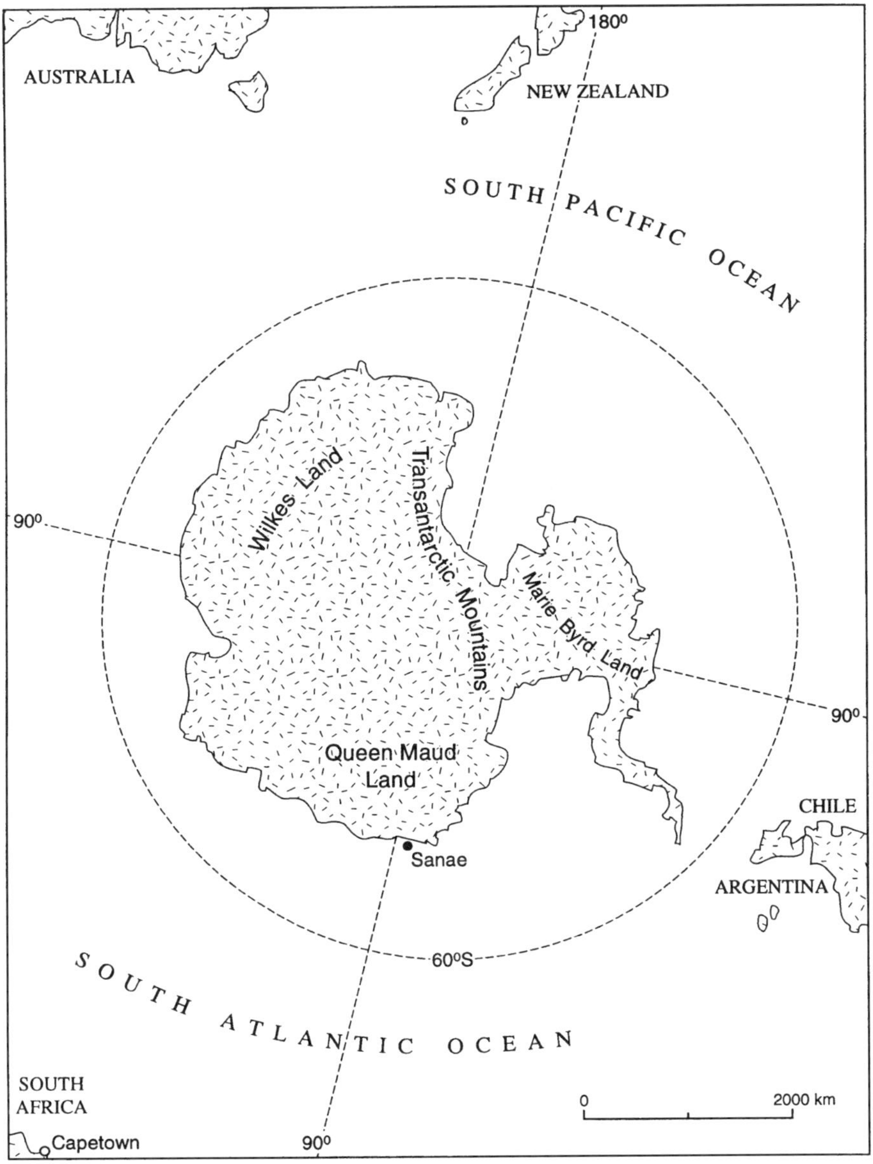

Map 8 South Africa's permanent base on the Antarctic continent

secure the environmental protection of the Antarctic are also well founded given concerns over ozone depletion and resource management. New opportunities to forge regional and southern hemispheric relationships, however, carry costs and responsibilities. South African investment in SANAE IV and the promotion of marine management in the Southern

Ocean will require investment in, *inter alia*, naval vessels, attendance at major international conventions and meetings, and support for scientific research. The growing importance of international environmental diplomacy has placed greater financial strains on South Africa, and in the future administrations will have to juggle competing national, regional and hemispheric interests and responsibilities. As the Deputy Minister of foreign affairs, Aziz Pahad, concluded at the meeting of the SAZOPC parties in July 1996:

> South Africa would recommend that at this meeting we focus on matters such as environmental pollution, managing the resources of the ocean, prevention of pollution, surveillance and sea rescue and general monitoring of the ocean . . . South Africa accepts the obligations to protect and preserve the marine environment as set out in the various conventions adopted internationally for this purpose in the past decade and more. We look forward to also pursue these principles within the Zone of Peace and Cooperation in the South Atlantic. We must protect this great body of water which divides us . . . but at the same time serves as the link between us. (cited in Mills 1996:9)

Summary

South Africa's involvement in the Antarctic and Southern Ocean has been a neglected dimension in many of the existing accounts of international relations and security policy. This chapter has explored the historical and political evolution of South African interests in the polar region. In the initial period of imperial polar policy, South Africa was the least significant of the three southern hemispheric Commonwealth nations. After the formal transfer of sovereignty of the Prince Edward Islands in 1948, South African activities have been fundamentally shaped by international reactions to apartheid and a strategic culture produced by the Cold War. Membership of the Antarctic Treaty System enabled South Africa to maintain a high level of diplomatic contacts throughout the post-war period of isolation and condemnation. The practical geopolitical reasoning of successive apartheid governments has concentrated on the strategic value of the region to South Africa in terms of resources, weather patterns, maritime access and geopolitical alliance-building. However, involvement in the ATS became increasingly precarious during the 1980s when controversies over common heritage and the management of the Antarctic became interlinked with widespread condemnation of South Africa. The participation of South Africa within the ATS subsequently became a useful political tool for countries such as Malaysia to use to attack the ATCPs more generally. The election of Nelson Mandela and the ending of apartheid in 1994 removed a festering source of tension within the ATS and the international community.

9

Conclusions and the future of Antarctica

Recalling his hazardous flight over the South Pole in 1929, Richard Byrd observed: "At the bottom of this planet is an enchanted continent in the sky, pale like a sleeping princess. Sinister – and beautiful – she lies in frozen slumber" (Byrd 1980[1938]:24). Today, Antarctica no longer lies in a frozen slumber since a range of human activities, above and below the frozen surface, has transformed our understanding of the polar continent. The recent discovery by Greenpeace scientists and members of the Argentine Antarctic Institute (AAI) of huge cracks in the Larsen B ice shelf has reminded the international community of the importance of that monitoring as it relates to the dangers of possible human-induced climatic change. Two years ago, the neighbouring Larsen A ice shelf collapsed and an iceberg, reportedly the size of Oxfordshire, floated towards the South Atlantic. Dr Rudi del Valle of the AAI was reported as stating that: "We saw a lot of cracks and ice rifts [in Larsen B] . . . and big holes in the ice shelf. And we don't have an explanation for them" (cited in *The Times*, 6 February 1997). Whilst there remains controversy, it is evident that the Antarctic will play a central role in the evaluation of climatic change and the formulation of responses to global environmental management.

For the Southern Oceanic Rim States (SORS), geographical proximity to the Antarctic and the Southern Ocean has been the key justification for considerable interest in the southern polar region's environmental and political management. "Effective stewardship" and "wise management" are the terms currently favoured by governments in the southern hemisphere anxious to ensure that limited financial resources are employed in the protection of the polar region. The recent debates in New Zealand over "Antarctic gateways" highlight an initial reconceptualization of the international politics of the Antarctic away from the alliances of the Cold War towards new sets of political and scientific relationships based on common geographical and environmental interests (Prior 1997). The success of new ventures such as the Valdivia Group in forging greater connections between the southern hemispheric consultative treaty parties will depend, *inter alia*, on the wider political management of the Antarctic region and common approaches to the implementation of the Environmental Protocol.

This final chapter reviews themes such as the settlement of the nuclear question in the Antarctic and the Southern Ocean and the environmental protection and treatment of territorial resource claims which have filled the agendas of foreign-policy managers in the southern hemisphere. Thereafter, the enduring issue of sovereignty, in particular with regard to the balancing of interests between claimant and non-claimant states, is considered. The future of the Antarctic Treaty System may ultimately depend upon this delicate balance in the midst of uncertainty over living and non-living resource agreements. Some final reflections on the future of Antarctica and the problems posed for those intent on its socio-political management are offered by way of conclusion (Stokke and Vidas 1996).

Antarctica and the Southern Oceanic Rim States

The historical and political evolution of the polar interests of a number of countries called the Southern Oceanic Rim States (SORS) have been investigated. The articulation of national interests have changed throughout this century, with the perceptions embodied by these governments strongly conditioned by their geographical location and geopolitical perspectives of the Antarctic and the Southern Ocean. Appeals to the importance of geographical proximity do not, however, in themselves provide sufficient ground for an explanation of the legal, resource, political and strategic interests of the SORS. Geographical proximity, whilst frequently commented upon by southern hemispheric political leaders, has to be understood within a wider historical and political context. In this sense, the broadly historical approach of this account has been designed to inform the reader of the evolution and development of these contexts. They include the articulation of national claims, the 1959 Antarctic Treaty, the debates over resource management and the recent concerns over the environmental protection of Antarctica.

This account has demonstrated that a number of common threads run through the political and historical evolution of the Antarctic interests of the SORS. The perceptual importance of geographical proximity has been one such theme. All the nations surveyed have argued, at times passionately, that their relative proximity to the Antarctic and the Southern Ocean implies that they have a pressing interest in shaping the political, strategic and environmental conditions of the region. The crux of the arguments have changed over time as during the 1950s, for example, when geographical factors were mobilized to argue that territorial claims might be compromised either by the activities of the Soviet Union and or by the collective Cold War designs of both superpowers. Popular media sources contributed by producing pessimistic strategic scenarios. In turn, these kinds of assessment encouraged political leaders to be dogmatic on the eve of important negotiations such as the 1959 Washington Conference. In the 1960s and 1970s, geographical factors were also frequently cited to explain why the SORS had to play a key role in shaping future conventions on resource management and conservation. During the 1980s and 1990s, the environmental protection of the Antarctic and the regulation of polar tourism emerged as the major themes which led the southern hemispheric nations to argue that their special interests had to be acknowledged too.

It has also been recognized that the SORS have, over the last 50 years, been anxious to ensure that the aforementioned interests were not compromised by the policies and designs of northern hemispheric states. In contrast to the Antarctic, the claimant states of the Arctic are all northern latitude and therefore geographically proximate nations. Moreover, both the United States and Russia are claimant states in the Arctic as opposed to their non-recognition of claims in the Antarctic region. The southern hemispheric nations have no direct political involvement in the Arctic region (although Chile is encouraging bipolar dialogue) and sometimes resent northern hemispheric involvement in the southern continent. This concern to limit the influence of northern states can be traced through a number of different contexts:

(a) nuclearization
(b) sovereignty and territorial claims
(c) the common heritage challenge of Malaysia
(d) development of environmental protection for the Antarctic.

The determination of the SORS to prevent the superpowers and Britain from conducting nuclear testing in the Antarctic was evident during the proceedings of the Washington Conference. Article V of the 1959 Antarctic Treaty was a diplomatic triumph for Argentina and the other SORS. Since the creation of the Antarctic Treaty System, the SORS have been zealous promoters of regional zones of peace and co-operation in all the oceans and seas of the southern hemisphere (Albuquerque 1989). The Declaration of the South Atlantic Zone of Peace and Co-operation in 1985, for example, elicited the following statement from President Raul Alfonsin of Argentina in September 1986:

> I am referring to the proposal to declare the South Atlantic a Zone of peace and co-operation for the peoples of South America and Africa, which was submitted by Brazil in the United Nations General Assembly and which has received the firm support of my country . . . The declaration of peace for the South Atlantic

> and its demilitarisation so as to avoid nuclearisation and the arms race in that fully corresponds to the policy of peace and disarmament by my government. (cited in Carasales 1989: 25)

The political context for this proposal was undoubtedly shaped by South American fears of an arms build-up either by Britain or an American-dominated NATO in the aftermath of the 1982 Falklands/Malvinas War and the onset of the Second Cold War. Whilst there remain real issues such as the implementation and management of a zone of peace and co-operation, it was indicative of a wider hemispheric concern that the activities of northern hemispheric nations such as Britain, France, the Soviet Union and the USA in these waters and others such as the Indian and Pacific Oceans should be curtailed. Jim Bolger's proposal in 1996 for a Southern Hemispheric Nuclear Free Zone underlines this concern for "regional" security.

The reluctance of the two superpowers either to declare a territorial claim in the Antarctic or to recognize the claims of the SORS created a considerable amount of political tension during the 1950s. In the aftermath of the Antarctic Treaty System, Article IV of the 1959 Antarctic Treaty placed territorial claims in a different legal and political context. Whilst these claims were effectively suspended, successive political elites of the SORS were determined that resource conventions and agreed measures for conservation recognize implicitly that they are geographically proximate claimant states. The living resource negotiations during the late 1970s saw the SORS concerns over ecological management and resource exploitation pitched against the major fishing interests of the Soviet Union and Japan, whilst the minerals resource convention in the 1980s recorded a degree of political tension between the USA and the SORS because of the former's insistence that its mineral interests should not be compromised by the concerns of the SORS over regulation and recognition of territorial sovereignty. The resource debates of the 1970s and 1980s reiterated this failure by the northern hemispheric states to recognize the claimant states of the SORS.

The debates over the future of the ATS and the Malaysian-led challenge in the UN General Assembly richly illustrated the common concerns of the SORS. As Richard Woolcott, the leading Australian representative, ironically noted in 1985: "Australia was surprised when Malaysia, a country of the northern hemisphere, launched its initiative for reasons which have never been entirely clear to many countries in the region" (cited in Beck 1994c: 23). The issues being raised by Malaysia were far broader than acknowledged by Richard Woolcott and other SORS representatives. Arvid Pardo, a LOS diplomat from Malta, had suggested in the 1960s that the deep seabed and its resources should be considered part of the common heritage of mankind rather than the exclusive property of a state (Pardo 1967, Glassner 1993). There were also widespread fears in the late 1970s that the forthcoming UNCLOS III conference would compromise SORS claims in the Antarctic region. The Chilean representative Fernando Zegers, for example, approached the New Zealand government in 1980 over a multilateral approach to this issue in the face of Third World pressure over common resource management and control of natural resources on the continental shelf and the deep seabed. The reactions of Zegers of Chile and Woolcott of Australia were emblematic of the hostility to northern hemispheric nations such as Malaysia or Sri Lanka either in questioning the political authority of the ATS and/or indirectly compromising their sovereign interests in the Antarctic and Southern Ocean (Joyner 1994, Vidas 1994).

The significance of these debates in the United Nations was enhanced further by the realization that the 1980s had been a "lost decade" in terms of economic development. The proposals to change the Antarctic into a place of common heritage related to established arguments over the need for a New International Economic Order (NIEO) (Herber 1991). The publication of *The Challenge to the South* by the South Commission in 1987 was indicative of this concern for economic justice, underdevelopment and the NIEO (Chatterjee and Finger 1995). Under the chairmanship of Julius Nyerere, the Commission concluded that there were three fundamental issues facing the Third World: promoting co-operation between the South, improving relations with the North and accurately diagnosing the economic and political problems faced by the South. The Commission proposed, *inter alia*, the creation of a Southern Bank and the enhanced transfer of technology and finance to the South from the North. The UN debates over the future of the Antarctic coincided with a dawning realization that the relatively developed nations of the SORS and the North generally were not prepared to renegotiate the international economic order. Unsurprisingly, given their national claims to the Antarctic, developing (and severely indebted) nations such as Argentina and Chile were unwilling to countenance a change in the political status of the Antarctic continent. This has prompted polar scholars such as Peter Beck to conclude that the UN debates over the "Question of Antarctica" have tended to be ineffectual because of the radically different starting positions of the participants (Beck 1986, 1994a, 1994c).

The development of environmental protection of the Antarctic necessitated a considerable amount of collective southern hemispheric diplomacy. The negotiation of the Protocol on Environmental Protection between 1989 and 1991 coincided with the growing "globalization" of environmental politics as demonstrated by the 1992 Earth Summit in Rio de Janeiro (Orrego Vicuna 1994). The political globalization of the Antarctic was a striking feature of the international politics of the 1980s and even led to fears within the ATS that the Earth Summit might propose to discuss the political future of the Antarctic. This was considered to be inappropriate by most members of the ATS because there was an existing international legal regime responsible for the polar continent. Political globalization in this sense, therefore, highlights the growing interdependence of states and the burgeoning role of non-state actors such as ASOC and Greenpeace in shaping international environmental politics (Walker 1988, Chatterjee and Finger 1995, Dalby 1996). In the case of inter-state co-operation, the creation of the Valdivia Group by a number of southern hemispheric nations in the mid-1990s was one response to the proliferation of international environmental measures and multilateral ecological diplomacy which undermined traditional United Nations groupings (see Hurrell 1994, 1995). As the opening declaration noted in 1995: "Encouraged by the principles and resolutions of the United Nations Conference on Environment and Development in June 1992, and following a period of consultations . . . representatives formed the Group of Temperate Southern Hemisphere Countries on Environment (Valdivia Group) to facilitate information exchange, identify positions and establish common courses of action and co-operation on international environment and related science issues" (New Zealand 1995). These recent developments have also touched upon the interests of other southern hemispheric nations with Antarctic interests such as Brazil and Uruguay. The former was admitted to the Valdivia Group in March 1996 and participated in the meeting of the Coordinating Committee in Buenos Aires in March 1997. Whilst it would be premature to evaluate the political effectiveness of the Valdivia Group, it is indicative of a concern throughout the southern hemisphere that collective environmental diplomacy, whether it be based on the Antarctic or ozone depletion, is vital in the face of the powerful northern hemispheric blocs of North America, the European Union as well as Japan.

The management of living resources and associated ecosystems within the Antarctic and Southern Ocean is a policy priority for the SORS. This can be illustrated by the accord signed between Argentina and New Zealand on 19 June 1996 pledging joint co-operation within the southern polar region. It includes such statements as: "Recognising the desirability of enhancing the development of the Antarctic Treaty System . . . Underlying the special importance that New Zealand and the Argentine Republic attach to the Antarctic and its associated and dependent ecosystems . . . Noting that New Zealand and the Argentine Republic are among the most important gateways to the Antarctic continent . . . Acknowledge their wish to explore the possibilities for co-ordinating their policies, where feasible, in international forums that deal with Antarctic matters" (Argentina/New Zealand 1996). It is increasingly likely that the so-called southern hemispheric "gateway" nations such as Argentina, Chile, Australia and New Zealand will seek to co-operate over the regulation of polar tourism and the environmental management of the region and that this will be an important agenda item for the 1997 ATCM in Christchurch (Prior 1997).

The management of living resources such as fish and whales is a common cause of concern for the SORS. The recent proceedings of the CCAMLR in Hobart have illustrated persistent mutual concerns over northern hemispheric fishing activities by nations such as South Korea, Japan, Spain and Russia. Fear of over-fishing in and around the waters of South Georgia and the South Shetlands, for example, have been further complicated by the ongoing sovereignty disputes between Argentina and Britain over the right of the latter to issue fishing licences and to declare an exclusive economic zone (Beck 1988). More generally, other South American nations have argued that the provisions of CCAMLR do not adequately recognize the legal and political problems relating to coastal jurisdiction and territorial sovereignty faced by Argentina and Chile.

Southern hemispheric nations have, therefore, particular geographical and political concerns relating to the Antarctic and Southern Ocean. The SORS will have to promote further diplomatic, environmental and scientific co-operation and recognize that northern hemispheric states such as the USA have their own vested interests in the development of environmental and resource conventions. New Zealand, South Africa and Chile, have initiated a dialogue about mutual environmental co-operation at the expense of national self-interest and territorial sovereignty. This does not imply, however, that territorial sovereignty and resource rights are declining in political saliency in these countries; rather it illustrates possibilities for more progressive forms of political co-operation. The enduring importance of resource issues within the Antarctic should not, therefore, be underestimated. As Francisco Orrego Vicuna concluded:

> There can be little doubt that minerals are available in Antarctica, and it is therefore entirely artificial to try and bury the question which can become alive the very minute a deposit is found. Should this be the case the Protocol would prove to be fundamentally unrealistic and its chances of survival would be virtually null . . . In the light of the above it does not seem to be entirely unwarranted to suggest that the whole matter should be thought over again . . . in particular to bring the minerals issue in line with both realism and traditions of the Antarctic Treaty System. (Orrego Vicuna 1994:8–9, see also Paltridge and Green 1995)

It remains a mute point as to whether this "realism" of the ATS is capable of transforming the central issue of competing claims to state sovereignty and resource rights.

Antarctic geopolitics, SORS and the sovereignty problematic

It has recently been argued that:

> The most significant area of continuity in Antarctic geopolitics relates to the array of territorial claims on the continent which, despite Article IV of the

> Antarctic Treaty, seems to have acquired new dimensions of complexity and contestation. The claimant states have not refrained from asserting sovereign rights over their claimed territories, be it in the way of issuing postage stamps, increased activity by national scientific expeditions, increasing their Antarctic budgets or the application of domestic or territorial laws to the claimed sector. (Chaturvedi 1996: 261)

In spite of measures such as the Protocol and the widespread commitment to the environmental protection of the Antarctic, the issue of sovereignty has not been resolved (Joyner 1994, Orrego Vicuna 1994). The centrality of sovereignty remains undiminished in spite of the multilateral co-operative framework developed since 1961 (Beck 1994c). The ATS has been interpreted by many political scientists and international lawyers as one of the most successful international regimes ever devised since the ending of the Second World War (Vidas 1994). Richard Falk states that the ATS was: "one of the few international arrangements that have succeeded . . . this is a legal arrangement that is reinforced by an appropriate geopolitical underpinning. It is not only that it has worked, but that the two superpowers are participating with the same understanding of their interest in keeping it working" (Falk 1986; cited in Kyriak 1992:119).

International politics have changed considerably since the comments about the ATS were made by Professor Falk in 1986. The ending of the Cold War in 1989 has provoked substantial change: "patterns of thought and behaviour identified as corresponding with an enduring, universal essence of global existence are coming under increasing scrutiny as old ideological commitments and alliances are reformulated, territorial boundaries are hastily redrawn, and new symbols of identity are constructed or resurrected" (George 1994:1). Ironically, whilst there were calls for the creation of a new world order in the 1990s, recent American administrations have illustrated a capacity for violent responses and big-power diplomacy which seem reminiscent of the Cold War rather than a new world order. American responses to the 1990–1 Gulf Crisis, the events in Somalia and Bosnia seem to exemplify this point. The ending of the Cold War also coincided with the negotiations over the Protocol on Environmental Protection for the Antarctic continent. Unlike the UN approach to the 1990–1 Gulf Crisis, the development of the Protocol was a progressive act in the sense that it recognized that mineral exploitation and resource rights had to be suspended in favour of a commitment to the environmental protection of the polar continent. Whilst environmental movements and some of the signatories to the Protocol have pointed to substantial legal and environmental flaws in the Madrid declaration, the immediate threat of mineral exploitation has been suspended for the near future (Stokke and Vidas 1996).

These observations are not intended to convey the impression that a new polar order is being constructed in the aftermath of the Protocol. There remains too much uncertainty over the implementation of the Protocol to be confident of a fundamental change to the international politics of the Antarctic. Several important issues need to be resolved by the ATS, including the following: the ban on mining is not permanent and therefore relies upon the self-restraint of the ATS states and third parties; the negotiation of the liability annexe remains unsettled, especially the delimitation of environmental liability; the execution and implementation of environmental impact assessments remains subject to variation because there are no universal guidelines for the Antarctic region; the treatment of marine pollution in the Antarctic is unclear given that naval warships and government vessels have been exempted from some of the provisions of the Protocol and the question of sovereignty has not been resolved. Whilst the Protocol has undoubtedly been considered a landmark agreement by most independent observers, as long as questions of liability, jursidiction and the ownership of resources such as ice are unsettled, tensions will endure.

This should not imply that the claimant states such as Argentina and Chile would be prepared to compromise their sovereign claims to the Antarctic in exchange for environmental protection. The rejection of CRAMRA and the development of the Protocol was only possible because Article 25 of the Protocol reaffirmed the principle of Article IV of the Antarctic Treaty regarding the respective rights of claimant and non-claimant states (Rothwell 1992, 1995). Although, international lawyers and political scientists have expounded much energy in examining the legal and political importance of sovereignty in the development of ATS, the existing territorial claims to the Antarctic have been a source of positive tension. Sovereignty issues have forced ATCPs to recognize that the ATS must be conserved and protected to allay fears of a return to the conflict-ridden years of the 1950s or the direct involvement of the UN. The case of Anglo-Argentine–Chilean relations is often considered emblematic of political accommodation and the strength of Article IV within the Antarctic Treaty. The Chilean writer Francisco Orrego Vicuna has suggested that:

> Sovereignty has become inseparable from co-operation in the Antarctic since it would be difficult to maintain sovereignty unless it was closely linked with the factor of co-operation. Conversely, it can also be said that co-operation on its own, if it does not rely on the support provided by national interest in the Antarctic, including those related to sovereignty, would probably lack the necessary vitality. (Orrego Vicuna 1988: 76)

Conversely, the persistence of these territorial claims in the Antarctic have provoked tensions within the ATS and even encouraged other states, via the United Nations, to raise the issue of ownership of the polar continent.

The four claimant states of the SORS are unlikely to relinquish their formal claims to the Antarctic in the near future. The Bolger government has been prepared to reconsider New Zealand's responsibilities in the Ross

Dependency in the aftermath of a major reorganization of their Antarctic programme and the newly created agency, Antarctica New Zealand. In 1992, New Zealand was the first government to conduct an independent audit of her national polar programme and in 1994 passed the Antarctica (Environment Protection) Act which stipulated the mechanisms to be introduced by New Zealand for the implementation of the Protocol. Environmental stewardship was introduced by the New Zealand Ministry of Foreign Affairs and Trade in 1994 to signify a more progressive relationship with the Ross Dependency rather than one based on narrow self-interest and territorial sovereignty. It was intended to convey that New Zealand had a special environmental and political relationship with the region which does not prohibit the promotion of international co-operation and effective management of Antarctic and Southern Ocean ecosystems. Evidence of this can be found in the creation of an International Centre for Antarctic Information and Research in 1991; the pooling of logistical resources with Italy and the USA; the creation of the Antarctic Heritage Trust in 1987; and the development of procedures for preliminary environmental assessment for all activities entering the Ross Dependency. In comparison with other SORS, New Zealand has adopted a more creative approach to the opportunities and constraints posed by territorial sovereignty within the Antarctic region, although at present the ideas associated with environmental stewardship remain poorly thought out in terms of their relationship to territorial sovereignty and political community. Notwithstanding these points, the changes instituted by the New Zealand government are a welcome intervention given the recent controversies over resource regulation and territorial sovereignty.

Final reflections on the future of Antarctica

In spite of the often radical divergence of opinions over the management of Antarctica, most commentators implicitly accept that the Antarctic is a special place of global importance. It is unusual because of its geographical isolation and lack of an indigenous population and with no single state in the position of representing the environmental, political and resource interests of the continent. The problem for the future management of the Antarctic or perhaps the problem of responding to the Antarctic continent through effective global action will continue to pose conceptual and practical difficulties (Hurrell 1995). Understanding these difficulties and acting to overcome them is often complicated by changes in the nature of international political orders and expanding geographical knowledge of the polar continent.

The political theorist, John Dunn once argued that the task of political theory is:

> Firstly . . . to train us in ascertaining how the social, political and economic setting of our lives now is and in understanding why it is as it is; secondly, in working through for ourselves how we could coherently and justifiably wish that world to be or become; and thirdly, in judging how far, and through what actions, and at what risk, we can realistically hope to move this world as it now stands towards the way we might excusably wish it to be. (Dunn 1990:193)

The elaboration of such tasks remains a problematic venture for political theory and other disciplines. Responding to the challenge posed by Dunn would require a careful examination not only of normative conceptions of how we respond to places such as the Antarctic but also an appreciation of practical political action (cf. Hurrell 1995:129, Vogler 1995). Within a political context, the recent debates over the implementation of the Protocol have highlighted the dilemma posed by the recognition that the environmental management of the Antarctic remains mediated by a fragmented system of sovereign states (Hurrell 1995:131–2). Whilst the ATS has been one of the most successful international regimes in recent political history, there is precious little agreement as to how the Protocol, which was designed to introduce environmental protection for the Antarctic, should be implemented, understood and evaluated.

The future management of the Antarctic will also depend on the evolving geographical and political understandings of that particular region. Ideas about place and the polar environment have not been static: over the present century, geographical knowledge of the polar continent and the Southern Ocean has increased significantly. In the era of remote sensing and satellite photography, it is easy to forget that knowledge of the Antarctic was fragmentary in the politically charged post-war period. Advanced technologies of exploration using the aeroplane and the aerial camera gave rise to new opportunities to explore, to press claims and to imagine the polar continent. As geographical knowledge of the Antarctic has expanded so the corpus of international law applied to the region has increased accordingly. In the 1950s, for example, the legal and geographical significance of ice, resources and the oceans was limited in comparison to pressing issues relating to land-based claims and the demilitarization of the continent. By the 1980s, however, the Law of the Sea negotiations coupled with ongoing research into resources and ocean management had transformed the Antarctic and the Southern Ocean (Watts 1992). The legal status of ice, for instance, remains uncertain because: "The international law concerning ice remains incomplete and unclear. No international regime is yet in place which comprehensively sets out the legal status of ice in its varying forms or specifically assigns jurisdictional competence over its use" (Joyner 1991 cited in Beck 1994c:25). This could become a major political and environmental issue if the harvesting of icebergs was to emerge as an important source of future water supply.

As geographical understanding of the Antarctic and the Southern Ocean has expanded, so different sets of political issues have emerged. For the first half of this century, the geopolitical destiny of the Antarctic was largely determined by a select group of European, American and Australasian nations. On the eve of the 1959 Washington Treaty, that group had expanded to include representatives from the African and Asian continents. Over the last 35 years, in spite of the impressive growth in membership of the ATS, the meaning and significance of the Antarctic has been tied up within the institutional and intellectual framework first developed by the original 12 states. The debates within the UN and social movements during the 1980s pointed to the fact that other actors were anxious to participate in the intellectual and political management of the Antarctic. Whilst it is likely that the political future of the Antarctic will continue to be fashioned primarily by the parties attached to the ATS, southern hemispheric nations will retain their important role in shaping resolutions concerning the management of tourism and the implementation of the Protocol. Perceptions of geographical proximity and overall geopolitical vision will undoubtedly continue to condition the policies and priorities of the SORS (Chaturvedi 1996:261). This study has indicated that the Antarctic and the Southern Ocean pose profound questions regarding resource management, territorial ownership, environmental protection, economic justice and development which are part of global politics rather than the narrow preserve of national or even hemispheric self-interest (Hurrell 1995, Volger 1995, Bretherton and Pontin 1996). The real challenge for the SORS and the international community is to build upon the positive legacies of the ATS and creatively to re-imagine the Antarctic and the Southern Ocean in a changing political world where the "fixed spaces" of international politics and political community are beginning to unravel.

References

Archival material

Australian Archives, Canberra. Records of the Department of External Affairs.

British Antarctic Survey, Cambridge. Records of the Falklands Islands Dependency Survey [BAS].

National Archives of Canada, Ottawa. Records of the Department of External Affairs [RG25].

National Archives of New Zealand, Christchurch and Wellington. Records of the Prime Minister's Department [PM] and the Department of Scientific and Industrial Research [DSIR].

Public Record Office, Kew (London). Records of the Admirality [ADM], Colonial Office [CO], Dominions Office [DO] and Foreign Office [FO].

Scott Polar Research Insititute, Cambridge. Records of the Antarctic Treaty Consultative Parties and the 1959 Washington Conference [MS 1308].

South African Archive Service, Pretoria. Records of the Department of Treasury [TES] and Commerce and Industries [HEN].

Documents of the Antarctic Treaty System

Antarctic Treaty Consultative Parties. 1975. *Final Report of the 7th Antarctic Treaty Consultative Parties* Oslo, 9–20 June 1975.

Antarctic Treaty Consultative Parties. 1977. *Final Report of the 9th Antarctic Treaty Consultative Parties* London, 19 September–7 October 1977.

Antarctic Treaty Consultative Parties. 1979. *Final Report of the 10th Antarctic Treaty Consultative Parties* Washington, 17 September–5 October 1979.

Antarctic Treaty Consultative Parties. 1981. *Final Report of the 11th Antarctic Treaty Consultative Parties* Buenos Aires, 23 June–7 July 1981.

Antarctic Treaty Consultative Parties. 1983. *Final Report of the 12th Antarctic Treaty Consultative Parties* Canberra, 13–27 September 1983.

Antarctic Treaty Consultative Parties. 1985. *Final Report of the 13th Antarctic Treaty Consultative Parties* Brussels, 7–18 October 1985.

Antarctic Treaty Consultative Parties. 1987. *Final Report of the 14th Antarctic Treaty Consultative Parties* Rio de Janeiro, 5–16 October 1987.

Antarctic Treaty Consultative Parties. 1989. *Final Report of the 15th Antarctic Treaty Consultative Parties* Paris, 9–19 October 1989.
Antarctic Treaty Consultative Parties. 1990. *Interim Report of the 11th Treaty Special Consultative Meeting* Vina del Mar, 19 November–6 December 1990.
Antarctic Treaty Consultative Parties. 1991. *Final Report of the 16th Antarctic Treaty Consultative Parties* Bonn, 7–18 October 1991.
Antarctic Treaty Consultative Parties. 1992. *Final Report of the 17th Antarctic Treaty Consultative Parties* Venice, 11–20 November 1992.
Antarctic Treaty Consultative Parties. 1994. *Final Report of the 18th Antarctic Treaty Consultative Parties* Kyoto, 11–22 April 1994.
Antarctic Treaty Consultative Parties. 1995. *Final Report of the 19th Antarctic Treaty Consultative Parties* Seoul, 8–19 May 1995.
Antarctic Treaty Consultative Parties. 1996. *Final Report of the 19th Antarctic Treaty Consultative Parties* Utrecht, 29 April–10 May 1996.

United Nations documents

UNGA. 1956. *United Nations General Assembly Records* A/3118/Addl. I 13 September 1956. A/3118/ Add.IIO 17 October 1956.
UNGA. 1958. *United Nations General Assembly Records* A/3852 15 July 1958.
UNGA. 1982. *United Nations General Assembly Records* A/37/PV 10. 29 September 1982.
UNGA. 1983. *United Nations General Assembly Records* A/C1/38/PV 46. 30 November 1983.
UNGA. 1984. *United Nations General Assembly Records*, 39th Session, Study on the Question of Antarctica, Views of States, A/35/583. Part II 29 October 1984.
UNGA. 1985. *United Nations General Assembly Records*, A40/C1/PV46. 17 November 1985.
UNGA. 1987. *United Nations General Assembly Records*, A42/C1/PV 46. 17 November 1987.
UNGA. 1988. *United Nations General Assembly Records*, A44/C1/PV 45. 22 November 1988.
UNGA. 1990. *United Nations General Assembly Records*, A45/C1/PV 40. 19 November 1990.
UN Secretariat 1984. *Question of Antarctica: Study Requested Under General Assembly Resolution 38/77, Report of the Secretary-General, Part Two (Views of States)*. Doc. A/39/583, Part II, 9 November 1984.

Official documents from the Southern Oceanic Rim States

Argentina, Comision Naçional del Antarctico. 1947. *Soberania Argentina en la Antartida* Buenos Aires: Comision Naçional del Antarctico.
Argentina, Comision Naçional del Antarctico. 1948. Las Islas Malvinas y el Sector Antarctico Argentina Buenos Aires: Comision Naçional del Antarctico.
Argentina, Comision Naçional del Antarctico. 1949. *Argentina Antartida* Buenos Aires: Comision Naçional del Antarctico.
Argentina, *Boletin Official* 19 March 1957.
Argentina, Direccion Naçional del Antarctico. 1988. *Tres Anos de Investigacion Antartica* Buenos Aires: Direccion Naçional del Antarctico.
Argentina, Insituto Geografico Militar. 1965. *Atlas de la Republica Argentina* Buenos Aires: IGM.

Argentina and New Zealand. 1996. *Joint Declaration of Antarctic Scientific Co-operation.*
Australia, Australian News and Information Bureau. 1961. *Australians in the Antarctic* Canberra: Australian News and Information Bureau.
Australia, Department of the Arts, Sport, the Environment, Tourism and Territories, 1988. *Report, Antarctic Science Advisory Committee* Canberra: Department of the Arts, Sport, the Environment, Tourism and Territories.
Australia, Australian Antarctic Division. 1980–1 *Annual Review* Hobart: Antarctic Division.
Australia, Australian Antarctic Division. 1991. *A Window on Antarctica* Hobart: Antarctic Division.
Australia, Australian Antarctic Division. 1995. *Looking South* Hobart: Antarctic Division.
Australia, Department of External Affairs. 1948. *Current Notes* Canberra: AGPS.
Australia, Department of Foreign Affairs and Trade. 1989–1992. *The Monthly Record* Canberra: AGPS.
Australia, Department of Foreign Affairs and Trade. 1985–1995. *Briefs for the Australian Delegations to the Antarctic Treaty Consultative Parties Conferences* Canberra: Department of Foreign Affairs and Trade.
Australia, Parliamentary Debates. 1948. Canberra: Commonwealth Government of Australia.
Australia, Parliamentary Paper. 1978. *Australia, Antarctica and the Law of the Sea* Canberra: Commonwealth of Australia.
India, Department of Ocean Development. 1982. *The First Indian Expedition to Antarctica December 1981–February 1982* New Delhi: Government of India.
India, Department of Ocean Development. 1984. *Annual Report 1983–1984* New Delhi: Government of India.
India, Department of Ocean Development. 1986a. *Annual Report 1985–1986* New Delhi: Government of India.
India, Department of Ocean Development. 1986b. *Third Indian Scientific Expedition to Antarctica. Technical Publication Number 3* New Delhi: Government of India.
India, Department of Ocean Development. 1987. *Fourth Indian Scientific Expedition to Antarctica. Technical Publication Number 4* New Delhi: Government of India.
India, Department of Ocean Development. 1988. *Annual Report 1987–1988* New Delhi: Government of India.
India, Department of Ocean Development. 1989. *Annual Report 1988–1989* New Delhi: Government of India.
India, Department of Ocean Development. 1990. *Report to SCAR 1 April 1989–31 March 1990 Indian National Science Academy* New Delhi: Government of India.
India, Department of Ocean Development. 1995. *Annual Report 1994–1995* New Delhi: Government of India.
India, Ministry of Information and Broadcasting. 1993. *India 1993: A Reference Manual* New Delhi: Government of India.
New Zealand, Ministry of Foreign Affairs and Trade. 1989. *White Paper on the Environment* Wellington: Government of New Zealand.
New Zealand, Department of Scientific and Industrial Research. 1983. *Antarctic Achievements 1957–1982: New Zealand's Role in the Antarctic* Christchurch: Government of New Zealand.
New Zealand, Ministry of Foreign Affairs and Trade. *Annual Report 1988–9* Wellington: Government of New Zealand.

New Zealand, Department of Foreign Affairs. 1995. *Final Report of the New Zealand Delegation to the CCAMLR 14th Meeting* Hobart: 23 October–5 November 1995.

New Zealand, Department of Foreign Affairs. 1996a. *Closing Statement for the Second Meeting of the Valdiva Group's Coordinating Committee* Canberra: 14–15 March 1996.

New Zealand, Department of Foreign Affairs. 1996b. *Antarctica and New Zealand* Information Bulletin Number 8 Department of Foreign Affairs, Government of New Zealand.

South Africa, Department of Environmental Affairs and Tourism. 1993. *Draft Comprehensive Environmental Evaluation of the Proposed new SANAE IV Facility at Vesleskarvet, Queen Maud Land, Antarctica* Pretoria: Republic of South Africa.

South Africa, Department of Environmental Affairs and Tourism. 1994. *Annual Report 1992–1993*. Pretoria: Republic of South Africa.

Books and articles

Acuna de Mones Ruiz, P. 1948. *Conciencia Antartica Argentina* Buenos Aires: Editorial Pleamar.

Agnew, J. 1993. "Representing space: space, scale and culture in social science" in J. Duncan and D. Ley eds *Place/Culture/Representation* London: Routledge.

Agnew, J. and Corbridge, S. 1989. "The new geopolitics: the dynamics of geopolitical disorder" in R.J. Johnston and P.J. Taylor eds *A World in Crisis? Geographical Perspectives* Oxford: Blackwell: 266–88.

Agnew, J. and Corbridge, S. 1995. *Mastering Space: Hegemony, Territory and International Political Economy* London: Routledge.

Akbar, M. 1985. *India: the Siege Within* Harmondsworth: Penguin.

Albuquerque, F. 1989. "Zona de paz y cooperacion en el Atlantico Sur" *Geopolitica* 38: 16–23.

Alley, R. 1984. "New Zealand and Antarctica" *International Journal* 34: 911–31.

Ambio, A Journal of the Human Environment 1989. XVIII(1). Special Issue on the Polar Regions.

Alzereca, C. 1949. *Historia de la Antartida* Buenos Aires: Editorial Hemisferio.

Anand, R. 1988. "Peace-time international co-operation for the security of the sea lanes of communication in the Indian Ocean" *International Studies* 25: 25–44.

Andersen, R. 1991. "Negotiating a new regime: how CRAMRA came into existence" in A. Jorgensen-Dahl and W. Ostreng eds *The Antarctic Treaty System in World Politics* London: Macmillan.

Anderson, M. 1996. *Frontiers* Oxford: Blackwell.

Anonymous. 1987. "An albatross around South Africa's neck" *New Scientist* 113 (1542): 26.

Antarctic and Southern Oceans Coalition. 1987. *Antarctica Briefing Number 12* 24 March 1987.

Archibugi, D. and Held, D. eds 1995. *Cosmopolitan Democracy* Cambridge: Polity.

Arenas, H. Santis. 1991. Personal communication regarding Canas Montalva 12 August 1991.

Arnello Romo, M. 1988a. "El mar y el futuro de Chile" *Revista Chilena de Geopolitica* 5(1): 33–7.

Arnello Romo, M. 1988b. "Chile: territorio y destino oceanico" *Chile: Historio y Geopolitico* 2(4): 52–4.

Arriagada, G. 1986. *Pinochet: The Politics of Power* Boston: Unwin Hyman.

Assef, A. 1980. *Proyeccion Continental de la Argentina* Buenos Aires: Pleamar.
ASOC Antarctic and Southern Ocean Coalition 1991. *Antarctic News* 9 November 1991.
Auburn, F. 1972. *The Ross Dependency* The Hague: Martin Nijhoff.
Auburn, F. 1982. *Antarctic Law and Politics* Bloomington: Indiana University Press.
Auburn, F. 1984. "Antarctic minerals and the Third World" *FRAM: A Journal of Polar Studies* 1(1): 200–23.
Auckland District Law Society. 1979. "How Strong is New Zealand's Claim to the Ross Dependency" Occasional Pamphlet Auckland: District Law Society.
Aylwin, P. 1990. "Principios y objectivos de la politica Internacional de Chile: Defensa Nacional y Fuerzas Armadas" *Geopolitica* 40: 7–15.
Azambuja, P. 1981. "Antartida: derecho que tiene Brazil" *Geosur* 23: 36–40.
Baachus, W. 1986. "*The South Atlantic War as a 'tip of the iceberg': expansionist interests in the Southern Atlantic region*" Unpublished paper University of Connecticut, Stamford.
Ball, D. 1980. *A Suitable Piece of Real Estate: American Installations in Australia* Sydney: Hale and Iremenger.
Bandyopadhyaya, J. 1979. "Dynamics of India's strategic environment" in B. Prasad ed. *India's Foreign Policy* New Delhi: Vikas Publishing: 21–34.
Barber, J. and Barratt, J. 1990. *South Africa's Foreign Policy* Cambridge: Cambridge University Press.
Barber, L. and Selby, M. 1984. "The search for an alternative strategy: New Zealand and the Antarctic" *The Round Table* 288: 466–72.
Barnes, J. and Nicholson, D. 1980. *The Leo Amery Diaries Volume 1* London: Hutchinson.
Barnes, T. and Ducan, J. eds *Writing Worlds: Discourse, Text and Metaphor in the Representation of Landscape* London: Routledge.
Bartolome, M. 1989. "Las relaciones Argentina–Brasil: del conflicto a la cooperacion" *Geopolitica* 39: 30–8.
Bartolome, M. 1991. "Conflictos en el Atlantico Sur en la decada del 90" *Geopolitica* 53: 56–7.
Barton, J. 1996. *Development, politics and space: towards a "southern" political geography* Research Papers in Environmental and Spatial Analysis, London School of Economics: Number 27.
Barton, J. 1997. *A Political Geography of Latin America* London: Routledge.
Bassin, M. 1987. "Race contra space: the conflict between German Geopolitik and National Socialism" *Political Geography Quarterly* 6 (2): 115–34.
Bassin, M. 1991. "Russia between Europe and Asia: the ideological construction of geographical space" *Slavic Review* 50: 1–17.
Bassin, M. 1994. "Russian geographers and the 'national mission' in the Far East" in D. Hooson ed. *Geography and National Identity* Oxford: Blackwell: 112–33.
Baughman, T. 1994. *Before the Heroes Came: Antarctica in the 1890s* Lincoln: University of Nebraska Press.
Beck, P. 1983. "Science – and politics – on ice" *Nature* 305: 107–8.
Beck, P. 1983–1984. "Securing the dominant place in the wan Antarctic sun for the British empire: the policy of extending control over Antarctica" *Australian Journal of Politics and History* 29.
Beck, P. 1984. "Britain and Antarctica: the historical perspective" *FRAM: Journal of Polar Studies* 1(1): 67–82.
Beck, P. 1985. "Preparatory meetings for the Antarctic Treaty 1958–59" *Polar Record* 22: 653–64.
Beck, P. 1986. *International Politics of Antarctica* New York: St Martins Press.

Beck, P. 1988. *The Falklands as an International Problem* London: Routledge.
Beck, P. 1989 "A Cold War" *Falklands Island Journal* pp. 36–43.
Beck, P. 1990a. "Antarctica as a zone of peace: A strategic irrelevance?" in R. Herr, H. Hall and M. Haward eds *Antarctica's Future: Continuity or Change?* Hobart: Tasmanian Government Printing Office: 193–223.
Beck, P. 1990b. "International relations in Antarctica: Argentina, Chile and the Great Powers" in M. Morris ed. *Great Power Relations in Argentina, Chile and Antarctica* London: Macmillan: 101–30.
Beck, P. 1990c. "Australia's new course in Antarctica" in W. Bateman and M. Ward eds *Australia's Maritime Interests* Canberra: Australian Centre for Maritime Studies.
Beck, P. 1994a. "Managing Antarctic tourism: a front-burner issue" *Annals of Tourism Research* 21(2): 375–86.
Beck, P. 1994b. "Asia, Antarctica and the United Nations" in R. Herr. and B. Davis eds *Asia and Antarctica* Canberra: Centre for Resource and Environmental Studies: 5–40.
Beck, P. 1994c. *Who Owns Antarctica? Governing and Managing the Last Continent* International Boundaries Research Unit, University of Durham.
Beck, U. 1992. *Risk Society* Cambridge: Polity.
Beeby, C. 1988 "The Convention on the Regulation of Antarctic Mineral Resource Activities" Speech to the International Bar Association 13 October 1988.
Beeby, C. 1972. *The Antarctic Treaty* Wellington: New Zealand Institute of International Affairs.
Beeby, C. 1991. "The Antarctic Treaty System: goals, performance and impact" in A. Morgensen-Dahl and W. Ostreng eds *The Antarctic Treaty System in World Politics* London: Macmillan.
Beltramino, J. 1993. *The Structure and Dynamics of Antarctic Populations* New York: Vantage Press.
Beltramino, J. ed. 1994. *Jornada Sobre Derecho y Geografia* Buenos Aires: Ediciones Manantial.
Beltramino, J. 1995. "Some considerations on the future of the Antarctic Treaty System" in A. Jackson ed. *On the Antarctic Horizon: Proceedings of the International Symposium on the Future of the Antarctic Treaty System* Hobart: Australian Antarctic Division 103–7.
Bennekom, S. van 1992. "A new regime to protect the Antarctic environment" *Leiden Journal of International Law* 5: 33–52.
Bergin, A. 1991. "The politics of Antarctic minerals: the greening of white Australia" *Australian Journal of Political Science* 26.
Berman, A. 1968. *South Africa and the Antarctic* Unpublished BA thesis University of Witswatersrand, Johannesburg.
Bhadauria, Y. 1983. "The Antarctic Treaty and its legal implications" *Indian Journal of International Law.*
Bierman, H. 1973. *The Republic of South Africa and the Southern Hemisphere* Potchefstroom: Centre for International Politics.
Biles, R. ed. 1991. *Inter-American Relations* Boulder: Lynne Rienner.
Black, J. 1986. *Sentinels of Empire: The United States and Latin America* New York: Greenwood.
Blij, H. 1978. "A regional geography of Antarctica and the southern oceans" *University of Miami Law Review* 33: 299–314.

Boczek, B. 1988. "The legal status of visitors, including tourists, and non-governmental organizations in Antarctica" in R. Wolfrum ed. *Antarctic Challenge III: Conflicting Interests, Environmental Protection, Economic Development* Berlin: Duncker and Humblot: 455–90.
Boletin Antarctico Chileno 1984. "Chile consolida a su soberania" 4: 59–62.
Bontori, O. 1988. *Brazil and the Vital South Atlantic* Washington DC: National Defense University Press.
Booth, K. ed. 1991. *New Thinking about Strategy and International Security* London: Macmillan.
Booth, K. and Vale, P. 1995. "Security in southern Africa: after apartheid, beyond realism" *International Affairs* 71: 285–304.
Borten, S. 1970. Editorial *Correo del Pinguino* 1(1): 1–2.
Bretherton, C. and Pontin, G. eds 1996. *Global Politics* Oxford: Blackwell.
British Antarctic Survey, 1990–1. *Annual Report* Cambridge: BAS.
Broadus, J. and Vartanov, R. eds 1995. *The Ocean and Environmental Security* Washington DC: Island Press.
Brown, A. 1995. "The design of CRAMRA: how appropriate for the protection of the environment?" *International Challenges* 10 (1) 24–6.
Bush, W. ed. 1982. *Antarctica and International Law* London: Oceana.
Bush, W. 1988. *Antarctica and International Law* New York: Oceana.
Bush, W. 1995. *Australia, Antarctica, the Minerals Convention and Environmental Protection* Antarctic and Southern Ocean Law and Policy Papers Number 7, University of Tasmania.
Bustamente, F. 1988. *La Antartida y el Pensamiento Geopolitico de las Fuerzas Armadas Chilenas* Santiago: FLASCO.
Buzan, B. 1991. *Peoples, States and Fear* Boulder: Lynne Rienner.
Byrd, R. 1980. *Alone* Washington DC: Island Press.
Campbell, D. 1992. *Writing Security* Manchester: Manchester University Press.
Campbell, D. 1993. *Politics Without Principle* Boulder: Lynne Rienner.
Campbell, D. 1995. "The new mind-set of international relations? Security, sovereignty and responsibility in *Cooperating for Peace*" in S. Lawson ed. *The New Agenda for Global Security: Cooperating for Peace and Beyond* Canberra: Allen and Unwin: 59–74.
Campbell, R. 1984. *Sea Power and South Africa* Pretoria: Institute of Strategic Studies: University of Pretoria.
Candioti, A. 1960. *El Tratado Antarctico y Nuestras Fuerzas Armadas* Buenos Aires: Pleamar.
Candioti, A. 1961. *El Tratado y el Derecho Publico Argentino* Buenos Aires: Pleamar.
Carasales, J. 1989. "The South Atlantic Zone of Peace and Co-operation" *World Review* 28(32): 22–50.
Casellas, A. 1974. *El Territorio Olivado* Buenos Aires: Insitituto de Publicaciones Navales.
Casellas, A. 1981. *Antartida: Un Malabarismo Politico* Buenos Aires: Editorial Pleamar.
Caviedes, C. 1988. "Fronteras, fronteras colonizables y fronteras geopoliticas en los paises del Cono Sur" in *International Symposium on Latin America* Varsovia: Poland: 45–65.
Charlton, M. 1989. *The Little Platoon* London: Macmillan.
Chatterjee, P. and Finger, M. 1995. *The Earth Brokers* London: Routledge.
Chaturvedi, S. 1983. "National claims on Antarctica" *Social Sciences Research Journal* VIII (1–2): 108–48.

Chaturvedi, S. 1985. "Antarctica and the Soviet Union" in R. Sharma ed. *Growing Focus on Antarctica* New Delhi: Rajesh Publications: 221–33.
Chaturvedi, S. 1986. "India and the Antarctic Treaty System: realities and prospects" *India Quarterly* 4: 1–38.
Chaturvedi, S. 1990. *The Dawning of Antarctica* New Delhi: Segment Books.
Chaturvedi, S. 1994. "*The Antarctic zone of peace*" Paper presented at the BISA Conference, University of York, 19–21 December 1994.
Chaturvedi, S. 1996. *The Polar Regions: A Political Geography* London: John Wiley and Sons.
Child, J. 1979. "Geopolitical thinking in Latin America" *Latin American Research Review* 14: 89–111.
Child, J. 1981. "Pensmiento geopolitica y cuatros conflictos en Sudamerica" *Revista de Ciencia Politica* 1–2: 71–104.
Child, J. 1985. *Geopolitics and Conflict in South America* New York: Praeger.
Child, J. 1986. "Latin lebensraum: the geopolitics of Ibero-American Antarctica" *Applied Geography* 10: 287–305.
Child, J. 1988. *Frozen Lebensraum: South American Geopolitics and Antarctica* New York: Praeger.
Child, J. 1989. "Antarctica: the South American geopolitical literature" *Inter-American Review of Bibliography* 39: 21–43.
Cirigliano, G. 1975. *La Argentina Triangular* Buenos Aires: Publications Humanitas.
Clapham, C. 1985. *Third World Politics: An Introduction* London: Croom Helm.
Cohen S. 1984. "Asymmetrical states and global geopolitical equilibrium" *SAIS Review* 4: 193–212.
Coker, C. 1982. "South Africa and the western alliance 1949–1981: a history of illusions" *RUSI Journal* 127: 34–40.
Conil Paz, A. and Ferrari, G. 1966. *Argentina's Foreign Policy 1930–1962* Notre Dame: University of Notre Dame Press.
Cook, G. ed. 1990. *The Future of Antarctica: Exploitation versus Preservation* Manchester: Manchester University Press.
Cooper, A. and Higgott, R. 1990/1. *Middle Power Leadership and Coalition Building: the Cairns Group and the Uruguay Round* Department of International Relations, Australian National University Working Paper Number 1990/1.
Cooper, J. and Headland, R. 1991. "A history of South African involvement in Antarctica and the Prince Edward Islands" *South African Journal of Antarctic Science* 21(2): 77–91.
Corbridge, S. 1993. "Colonialism, post-colonialism and the political geography of the Third World" in P. Taylor ed. *Political Geography of the Twentieth Century* London: Belhaven Press: 171–206.
Corbridge, S., Martin, R. and Thrift, N. eds 1995. *Money, Power and Space* Oxford: Blackwell.
Cortese, E. 1981. "La Argentina: pais geopoliticamente agredido" *Geopolitica* 21: 47–50.
Cosentino, B. 1970. "El valor estrategico de las Malvinas" *Estrategia* 6: 76–87.
Cosgrove, D. 1994. "Contested global visions: one world, whole earth and the Apollo space photographs" *Annals of the Association of American Geographers* 84(2): 270–94.
Couper, A. 1978. *Geography and the Law of the Sea* London: Macmillan.
Cox, M. 1990. "From the Truman doctrine to the second superpower debate: the rise and fall of the Cold War" *Journal of Peace Research* 27(1): 25–41.

Crawford, A. 1950. "The establishment of the South African meteorological station on Marion Island" *Polar Record* 5: 576–9.
Crawley, E. 1984. *A House Divided* London: I.B. Tauris.
Cresswell, T. 1996. *In Place/Out of Place: Geography, Ideology and Transgression* Minneapolis: University of Minnesota Press.
Cromley, R. 1958. "We're losing the Antarctic" *American Mercury* 87: 5–11.
Cross, M. 1991. "Antarctica: exploration or exploitation?" *New Scientist* 22 June 1991 25–8.
Cumpston, M. 1995 *History of Australian Foreign Policy 1901–1991* Canberra: Union Offset.
Cunningham, J. 1958. "A politico-geographical appreciation of New Zealand foreign policy" *New Zealand Geographer* 14: 147–60.
Dalby, S. 1990. *Creating the Second Cold War* London: Pinter.
Dalby, S. 1991. "Critical geopolitics: discourse, difference and dissent" *Environment and Planning D: Society and Space* 9: 261–83.
Dalby, S. 1993. "The Kiwi disease: geopolitical discourse in Aotearoa/New Zealand and the South Pacific" *Political Geography* 12: 437–56.
Dalby, S. 1994. *Contesting an essential concept: dilemmas in contemporary security discourse* Occasional Papers Number 6. Ottawa: The Norman Patterson School of International Affairs, Carleton University.
Dalby, S. 1996. "Reading Rio, writing the world: the New York Times and the Earth Summit" *Political Geography* 15(6/7): 593–614.
Dalby, S. and Ó Tuathail, G. 1996. "Editorial Introduction: the critical geopolitics constellation: problematising fusions of geographical knowledge and power" *Political Geography* 15(6/7): 451–6.
Dalziell, J. 1997. "A new focus for Greenpeace" *Polar Record* 33: 3–4.
Darby, A. 1994. *Non-State Actors in the Antarctic Treaty System* Hobart: Antarctic CRC Research Report 4.
Daus, F. 1950. *Geografia en Argentina* Buenos Aires: Editorial Pleamar.
Davies, R. 1995. *South Africa's Foreign Policy Options in a Changing Global Context* Centre for Southern African Studies: University of Western Cape.
Davis, B. 1992a. "Antarctica as a global protected area: perceptions and reality" *Australian Geographer* 23: 39–43.
Davis, B. 1992b. "Focussing an Antarctic research programme: the Australian experience" *Polar Record* 28: 51–6.
Davis, B. 1995. "Re-inventing offshore Australia: oceans policy in the post-UNCED era" *Australian Geographer* 26: 38–44.
Denman, D. 1990. "Privatise Antarctica" *Economic Affairs* October/November 27–9.
De Castro, T. 1976. *Rumo a Antartica* Rio de Janeiro: Freitas.
De Hoyos, R. 1988. "Malvinas/Falklands, 1982–1988: the new Gibraltar in the South Atlantic?" in P. Kelly and J. Child eds *Geopolitics of the Southern Cone and Antarctica* Boulder: Lynne Rienner: 237–49.
De Wit, M. 1985. "Reshaping the Antarctic Treaty, with implications for South Africa" *International Affairs Bulletin* 9(2): 44–58.
Dey, A. 1992. "India in Antarctica – Perspectives, programmes and achievements" *International Studies* 29: 173–85.
Dey-Nuttall, A. 1994. *Origins, Development and Organization of National Antarctic Programmes: With Special Reference to the United Kingdom and India* Unpublished PhD thesis Cambridge, University of Cambridge.
Dieguez, M. 1989. "La neutralidad de Argentina durante la Segunda Guerra" *Estudios Internacionales* 22: 53–77.

Dodds, K. 1993. "Geopolitics, cartography and the state in South America" *Political Geography* 12: 361–81.
Dodds, K. 1994a. "Creating a strategic crisis out of a communist drama: Argentine and South African geo-graphs of the South Atlantic" *European Review of Latin American and Caribbean Studies* 56(1): 33–55.
Dodds, K 1994b. "Geopolitics in the Foreign Office: British representations of Argentina 1945–1961" *Transactions of the Institute of British Geographers* 19: 273–91.
Dodds, K. 1995. "South Africa, the South Atlantic and the international politics of Antarctica" *South African Journal of International Affairs* 3: 60–80.
Dodds, K. 1996a. "South Africa and the Antarctic, 1920–1960" *Polar Record* 32(180): 25–42.
Dodds, K. 1996b. "To photograph the Antarctic: British polar exploration and the FIDAS expedition" *Ecumene* 3(1): 63–89.
Dodds, K. 1996c. "The end of a polar empire? The Falkland Islands Dependencies and Commonwealth reactions to British polar policy" *Journal of Imperial and Commonwealth History* 24: 392–421.
Dodds, K. 1997. "Antarctica and the modern geographical imagination 1918–1960" *Polar Record* 33(184): 47–62.
Dodds, K. and Sidaway, J. 1994. "Locating critical geopolitics" *Environment and Planning D: Society and Space* 12: 515–524.
Dorrance, J. 1985. "ANZUS misperceptions, mythology and reality" *Australian Quarterly* Spring: 215–30.
Duncan, J. and Ley, D. eds 1992. *Place/Culture/Representation* London: Routledge.
Dunn, J. 1990. *Interpreting Political Responsibility* Cambridge: Cambridge University Press.
Dunn, R. 1989. "Australia to seek wilderness park status in Antarctica" *Financial Review* 23 May 1989.
Elliot, D. 1988. "Antarctica: is there any oil and gas?" *Oceanus* 31(2): 32–8.
Elliot, L. 1994a. *International Environmental Politics: Protecting the Antarctic* New York: St Martin's Press.
Elliot, L. 1994b. "Environmental protection in the Antarctic: problems and prospects" *Environmental Politics* 3: 247–72.
Enzenbacher, D. 1992. "Tourists in Antarctica: numbers and trends" *Polar Record* 28: 17–22.
Escobar, A. 1995. *Encountering Development* Princeton: Princeton University Press.
Escolar, M., Quintero Palacios, S. and Reobratti, C. 1994. "Geographical identity and patriotic education in Argentina" in D. Hooson ed. *Geography and National Identity* Oxford: Blackwell: 346–66.
Escudé, C. 1983. *Estados Unidos y la Declinacion Argentina 1942–1949* Buenos Aires: Editorial Pleamar.
Escudé, C. 1984. *La Argentina: Paria Internacional?* Buenos Aires: Editorial del Belgrano.
Escudé, C. 1986. "Contenido nacionalista de la ensenanza de la geografia en la Republica Argentina 1879–1986" *Ideas en Ciencias Sociales* 9: 3–43.
Escudé, C. 1987. *Patalogia del Nacionalismo: el caso Argentina* Buenos Aires: Instituto de Torcuato Di Tella.
Escudé, C. 1988. "Argentine territorial nationalism" *Journal of Latin American Studies* 20: 139–65.
Escudé, C. 1990. *El Fracaso del Proyecto Argentino* Buenos Aires: Instituto de Torcuato Di Tella

Escudé, C. 1992. *Education, political culture and foreign policy: the case of Argentina* Working Paper Series of the Duke–UNC Program of Latin American Studies.

Ferguson, R. 1995. *Antarctic Artefacts: A Selective Description and Annotated Catalogue of Artefacts in The Mawson Collection* Adelaide: University of Adelaide Press.

Fogg, G. 1992. *A History of Antarctic Science* Cambridge: Cambridge University Press.

Forrest, T. 1981. "Brazil and Africa: geopolitics, trade and technology in the South Atlantic" *African Affairs* 81: 3–21.

Foucault, M. 1984. "The order of discourse" in M. Shapiro ed. *Language and Politics* Oxford: Basil Blackwell.

Fox, A. 1980. "The range of choice for middle powers: Australia and Canada compared" *Australian Journal of Politics and History* 26(2): 193–203.

Fraga, J. 1979. *Introduccion a la Geopolitica Antarctica* Buenos Aires: Direccion Nacional del Antarctico.

Fraga, J. 1983. *Geopolitica y el Atlantico Sur* Buenos Aires: Pleamar.

Fuchs, V. and Hillary, E. 1958. *The Crossing of Antarctica* London: Cassell.

Gajardo Villarroel, E. 1983. "Chile, el tratado Antarctico y su sistema" *Boletin Antarctico Chileno* 3: 7–15.

Garcia Monge, M. 1990–1. 'Antarctico del Siglo XXI: Un sistema cooperacion ejemplar tarea para las nuevas generaciones" *Boletin Antarctico Chileno* 10: 27–41.

Geldenhuys, D. 1981. *Some Foreign policy Implications of South Africa's "Total National Strategy" with Particular Reference to the 12 Point Plan* Johannesburg: SAIIA.

Geldenhuys, D. 1984. *The Diplomacy of Isolation* Johannesburg: Macmillan.

Geldenhuys, D. 1990. "A comprehensive analysis of the international isolation of Pinochet's Chile and South Africa under nationalist rule" *UNISA* 6(1): 18–27.

Genest, E. 1985. *Naturaleza y Alcances de nuestra Reinvindicacion Territorial Sobre Tierra e Islas Polares* Buenos Aires, Instituto Antartico Argentina.

Genest, E. 1992. *Las Organzaciones no gubernamentales y el Sistema del Trado Antarctico* Buenos Aires: Direccion Nacional del Antarctico.

George, J. 1994. *Discourse of Global Politics* Boulder: Lynne Rienner.

Glassner, M. 1985. "The View from the near south: South Americans view Antarctica and the Southern Ocean geopolitically" *Political Geography Quarterly* 4: 329–42.

Glassner, M. 1990. *Neptune's Domain* London: Allen and Unwin.

Glassner, M. 1993. *Political Geography* Chichester: John Wiley.

Glassner, M. 1995. *Political Geography* Chichester: John Wiley.

Godlewska, A. and Smith, N. eds 1994. *Geography and Empire* Oxford: Blackwell.

Gough, B. 1992. *The Falklands/Malvinas: The Conquest for Empire in the South Atlantic* London: Athlone Press.

Gould, L. 1958. *The Polar Regions in Their Relation to Human Affairs* New York: American Geographical Society.

Gravil, R. 1985. *The Anglo-Argentine Connection 1900–1939* Boulder: Westview.

Gravil, R. 1989. "The Southern Cone and South Africa" *International Affairs Bulletin* 13(2): 26–36.

Grolin, J. 1987. "The question of Antarctica and the problem of sovereignty" *International Relations* 9: 39–55.

Gugliamelli, J. 1976. "Geopolitica en la Argentina" *Estrategia* 46–7: 5–14.

Gustafson, L. 1988. *The Sovereignty Dispute over the Falklands/Malvinas Islands* Oxford: Oxford University Press.

Guyer, R. 1983. "Antarctica's role in international relations" in F. Orrego Vicuna ed. *Antarctic Resources Policy: scientific, legal and political issues* Cambridge: Cambridge University Press: 267–79.

Gygli, P. 1976. "Power-political position and strategic problems of the Republic of South Africa" *South African International* 6: 109–16.

Haggman, B. 1994. *Geopolitics in India* Helsingborg: Centre for the Study of Geopolitics Research Paper 17.

Hall, C. and Johnston, M. ed. 1995. *Polar Tourism: Tourism in the Arctic and Antarctic Regions* Chichester: John Wiley.

Hall, R. 1986. *Antarctica and World Politics: The Significance of Political Factors in Antarctic Affairs During the 20th Century* Unpublished MA thesis, University of Tasmania Hobart.

Hanessian, J. 1965. "The Antarctic Treaty 1959" *International and Comparative Law Quarterly* 9: 436–80.

Hanlon, J. 1986. *Beggar Your Neighbour* Bloomington: University of Indiana Press.

Harley, J. 1988. "Maps, knowledge and power" in D. Cosgrove and S. Daniels eds *Iconography of Landscape* Cambridge: Cambridge University Press.

Harley, J. 1992. "Deconstructing the map" in T. Barnes and J. Duncan eds *Writing Worlds* London: Routledge.

Harris, W. 1991. "New Zealand defence policy" *New Zealand Geographer* 47: 72–9.

Hayes, J. 1928. *Antarctica: a Treatise on the Southern Continent* London: Richard's Press.

Headland, R. 1985. *South Georgia* Cambridge: Cambridge University Press.

Headland, R. 1989. *Chronological List of Antarctic Expeditions and Related Historical Events* Cambridge: Cambridge University Press.

Heap, J. 1983. "Co-operation in the Antarctic: a quarter of a century's experience" in F. Vicuna eds *Antarctic Resources Policy: Scientific, Legal and Policy Issues* Cambridge: Cambridge University Press: 103–8.

Heap, J. 1988. "The role of scientific advice for the decision making process in the Antarctic Treaty system" in R. Wolfrum and E. Michos-Ederer eds *Antarctic Challenge III* Berlin: Duncker and Humblot: 21–8.

Henshaw, P. 1992. "The transfer of Simonstown: Afrikaner nationalism, South African strategic dependence and British global power" *Journal of Imperial and Commonwealth History* 20: 419–44.

Hepple, L. 1986. "The revival of geopolitics" *Political Geography Quarterly* 5(4): 21–36.

Hepple, L. 1988. "The geopolitics of the Falklands/Malvinas and the South Atlantic: British and Argentina perceptions, misperceptions and rivalries" in P. Kelly and J. Child eds *Geopolitics of the Southern Cone and Antarctica* Boulder: Westview: 250–82.

Hepple, L. 1992. "Metaphor, geopolitical discourse and the military in South America" in T. Barnes and J. Duncan eds *Writing Worlds* London: Routledge.

Herber, B. 1991. "The common heritage principle: Antarctica and the developing nations" *American Journal of Economics and Sociology* 50: 391–406.

Heron, D. 1954. "Antarctica claims" *Foreign Office* 32: 661–7.

Herr, R. 1984. "The changing geo-politics of ANZUS: the place of the South Pacific" *World Review* 23: 21–42.

Herr, R. and Davis, B. eds 1994. *Asia in Antarctica* Hobart: Antarctic Cooperative Research Centre.

Herr, R. and Davis, B. 1996. "ATS decision making and change: the role of domestic politics in Australia" in D. Vidas and O. Stokke eds *Governing the Antarctic* Cambridge: Cambridge University Press.
Herr, R. and Hall, R. 1990. "Science as currency and the currency of science" in R. Herr, R. Hall and M. Hayward eds *Antarctica's Future: Continuity or Change?* Hobart: Government Printing Office.
Hewitt, V. 1992. *The International Politics of South Asia* Manchester: Manchester University Press.
Heymann, G. et al 1987. *Report to the Minister of Environmental Affairs on an Environmental Impact Assessment of a Proposed Emergency Landing Facility on Marion Island* Pretoria: South African National Scientific Programmes Report 140.
Hill, J. 1983. *New Zealand and Antarctica: a discussion paper* Wellington: Ministry of Foreign Affairs.
Holbraad, C. 1984. *Middle Powers in International Politics* London: Macmillan.
Holdar, S. 1992. "The ideal state and the power of geography: the life and work of Rudolf Kjellen" *Political Geography* 11: 307–23.
Holdgate, M. 1987. "Regulated development and conservation of Antarctic resources" in G. Triggs ed. *The Antarctic Treaty Regime: Law, Environment and Resources* Cambridge: Cambridge University Press: 128–42.
Hopkins, T. and Wallerstein, I. 1982. *World-Systems Analysis* London: Sage.
Hooson, D. ed. 1994. *Geography and National Identity* Oxford: Blackwell.
Hunter Christie, E. 1951. *The Antarctic Problem* London: George Allen and Unwin.
Hurrell, A. 1983. "The politics of South Atlantic security" *International Affairs* 59: 179–93.
Hurrell, A. 1994. "A crisis of ecological viability? global environmental change and the nation state" *Political Studies* XLII: 146–65.
Hurrell, A. 1995. "International politics theory and the global environment" in K. Booth and S. Smith eds *International Relations Today* Cambridge: Polity: 129–53.
Ihl, P. 1952. "Linea geopolitica de Chile" *Revista Geografia de Chile* 8: 26–43.
Illingworth, F. 1953. "Scramble for Antarctica" *New Commonwealth* 23 November 1953.
Infante, M. 1993. *Chilean Antarctic Policy: The Influence of Domestic and Foreign Policy* IARP Publication Series Number 2 Oslo: Fridtjof Nansen Institute.
International Service of Argentine Publications 1954. *The Argentine Antarctic* Buenos Aires: ISAP.
Isola, E. and Berra, A. 1950. *Introduccion a la Geopolitica Argentina* Buenos Aires: Editorial Pleamar.
Jacka, F. and Jacka, E. eds 1988. *Mawson's Antarctic Diaries* Sydney: Allen and Unwin.
Jalonen, O. 1988. "The strategic significance of the Arctic" in K. Mottola ed. *The Arctic Challenge: Nordic and Canadian Approaches to Security and Cooperation in an Emerging International Region* Boulder: Westview: 157–82.
Jilberto, A. 1991. "Military bureaucracy, political opposition and democratic transition" *Latin American Perspectives* 18(1): 33–65.
Johnston, M. 1997. "Polar regulation strategies: controlling vistors through codes of conduct and legislation" *Polar Record* 33(184): 13–20.
Joyner, C. 1984. "Anglo-Argentine rivalry after the Falklands/Malvinas War: law, geopolitics and the Antarctic connection" *Lawyer of the Americas* 15: 467–502.
Joyner, C. 1985. "Polar politics in the 1980s: some preliminary thoughts on polar contrasts and geopolitical considerations" *International Studies Notes* 11: 1–5.

Joyner, C. 1990a. "Antarctica and the Indian Ocean States: the interplay of law, interests and geopolitics" *Ocean Development and International Law* 21: 41–70.
Joyner, C. 1990b. "Maritime zones in the southern oceans: problems concerning the correspondence of natural and legal martime zones" *Applied Geography* 10: 307–25.
Joyner, C. 1992. *Antarctica and the Law of the Sea* The Hague: Martinus Nijhoff.
Joyner, C. 1994. *The Legitimacy of the Convention on the Regulation of Antarctic Mineral Resource Activities* IARP Publication Series Number 3 Oslo: Fridtjof Nansen Institute.
Joyner, C. 1995. *CRAMRA's legacy of legitimacy: progenitor to the Madrid environmental protocol* Antarctic and Southern Ocean Law and Policy Occasional Papers Number 7 University of Tasmania.
Kato, M. 1993. "Nuclear globalism: traversing rockets, satellites and nuclear war via the strategic gaze" *Alternatives* 18: 339–60.
Kearns, G. 1984. "Closed space and political practice: Frederick Jackson Turner and Halford Mackinder" *Environment and Planning D: Society and Space* 1: 23–34.
Kelly, P. 1985. "*Geopolitics as a foreign policy approach: the case of US relations towards Latin America*" Paper presented to the Southwestern Political Science Association, Houston, Texas, 21–3 March 1985.
Kelly, P. 1996. *South American Geopolitics* Austin: University of Texas Press.
Kelly, P. and Child, J. eds 1988. *Geopolitics of the Southern Cone and Antarctica* Boulder: Lynne Rienner.
Kemp, G. 1977. "The new strategic map" *Survival* 19(2): 50–9.
Kennedy, P. 1988. *The Rise and Fall of Great Powers* Harmondsworth: Penguin.
King, J. 1952. "South Africa and the sub-Antarctic" in F. Simpson ed. *The Antarctic Today* Wellington: AH and AW Reed: 304–12.
Kingwill, D. 1971. "First ten years of South African Antarctic research" *South African Journal of Antarctic Research* 1(1): 2–3.
Klein, B. 1994. *World Order and Strategic Studies* Cambridge: Cambridge University Press.
Kliot, N. and Waterman, S. eds 1992. *The Political Geography of Peace and Conflict* London: Pinter.
Klotz, F. 1990. *America on the Ice: American Policy Issues* Washington DC: National Defense University Press.
Kyriak, M. 1992. "The future of the Antarctic Treaty System: an examination and evaluation of the 'common heritage' and the 'world park' proposals for an alternative regime" *Auckland University Law Review* 7: 105–26.
Kyriak, M. 1996. *New Zealand's Antarctic Policy 1988–1991* Unpublished PhD Thesis, University of Auckland.
Labra, O. 1947. *Chilenos en la Antartida* Santiago: Andres Bello.
Lacoste, Y. 1976. *La Geographie ca sert d'abord à faire la Guerre* Paris: Maspero.
Lacoste, Y. 1984. "Geography and foreign policy" *SAIS Review* 4: 213–27.
La Grange, J. 1991. "Taking part in the Trans-Antarctic Expedition" *South African Journal of Antarctic Research* 21: 98–106.
Larus, J. 1981. "India's non-alignment and superpower rivalry in the Indian Ocean" in L. Bowman and I. Clark eds *The Indian Ocean in Global Politics* Boulder: Westview: 41–57.
Larus. J, 1984. "India claims a role in Antarctica" *The Round Table* 289: 45–56.
Latour, B. 1987: *Science in Action* Milton Keynes: Open University Press.

Laverde, R. 1988. *The Development, Pursuit and Maintenance of a South African Antarctic policy, 1926–1988* Unpublished MA thesis Grahamstown, Rhodes University.
Law, P. 1983. *Antarctic Odyssey* Melbourne: Heinemann.
Laws, R. 1987. "Science, the Treaty and the future" in D. Walton ed. *Antarctic Science* Cambridge: Cambridge University Press: 250–65.
Lee, D. 1995. *Search for Security: the Political Economy of Australia's Foreign and Defence Policy* Canberra: Allen and Unwin.
Legu'be, J. 1974. *Cap Sud: le Pivot Africain* Paris: Editions Lavauzelle.
Lemon, A. ed. 1995. *The Geography of Change in South Africa* London: John Wiley.
Lewis, R. 1965. *A Continent for Science* London: Secker and Warburg.
Lewis, R. and Smith, P. eds 1973. *Frozen Future* New York: McGraw-Hill.
Leysens, A. 1990. "South Africa's international relations with Latin America: an appraisal" *International Affairs Bulletin* 14(1): 33–43.
Livingstone, D. 1992. *The Geographical Tradition* Oxford: Blackwell.
Logan, H. 1979. *Cold Commitment: the Development of New Zealand's Territorial Role in Antarctica 1920–1960* Unpublished MA thesis Christchurch, University of Canterbury.
Lopez, B. 1986. *Arctic Dreams: Imagination and Desire in a Northern Landscape* London: Macmillan.
Lopez, B. 1988. "Trouble way down under – exploiters threaten the eerie beauty of Antarctica" *Washington Post* 27 March 1988.
Lovering, J. and Prescott, V. 1979. *Last of Lands . . . Antarctica* Melbourne: University of Melbourne Press.
Luard, E. 1984. "Who owns the Antarctic" *Foreign Affairs* 52: 1174–93.
Luke, T. 1996. "Governmentality and contragovernmentality: rethinking sovereignty and territoriality after the Cold War" *Political Geography* 15(6/7): 491–508.
Lyons, D. 1993. "Environmental impact assessment in Antarctica under the Protocol on Environmental Protection" *Polar Record* 29: 111–20.
Mann, M. 1993. *Sources of Social Power* vol. 2 Cambridge: Polity.
Manzoni, M. and Pagini, P. 1994. "Potere e strategie: aspetti di geografia politica Antarctica" *Memoirie: Societa Geografica Italiana* 51: 185–208.
Manzoni, M. and Pagini, P. 1996. "The symbolic territory of Antarctica" *Political Geography* 15: 359–64.
Marcella, G. 1990. "The Latin American military, low-intensity conflict and democracy" *Journal of Interamerican Studies* 32(1): 45–83.
Marini, R. 1972. "Brazilian sub-imperialism" *Monthly Review* 23(9): 14–24.
Marsh, J. 1948. *No Pathway Here* Cape Town: Howard Timmins.
Mason, J. 1991. *The Cold War 1945–1991* London: Routledge.
Mawson, D. 1911. "The Australian Antarctic Expedition" *Geographical Journal* 37: 609–20.
Mawson, D. 1914. "The Australian Antarctic Expedition 1911–14" *Geographical Journal* 44: 257–86.
McCannon, J. 1994. *Red Arctic: The political and cultural significance of the Arctic in the Soviet Union 1932–1939* Unpublished PhD thesis, University of Chicago.
McGrew, A. 1992. "Conceptualising world politics" in A. McGrew and P. Lewis eds *Global Politics* Cambridge: Polity: 1–30.
Melinsky, E. 1978. *Argentina's Foreign Policies* Boulder: Westview.
Mercado Jarrin, E. 1984. *Peru y la Antartida* Lima: Instituto Peruano de Estudios Geopoliticos y Estrategicos.

Mericq, L. 1988. *Chile's Antarctic Claim* Washington DC: National Defense University Press.
Merino Castro, J. 1987. "La Antartida y el Futuro de Chile" *Boletino Antartico Chileno* 7(1): 27–41.
Milia, F. ed. 1978. *Atlantarida* Buenos Aries: Editorial Pleamar.
Mill, H. 1896. "The International Geographical Congress" *Geographical Journal* 8: 292.
Mill, H. 1901. *The Siege of the South Pole* London: Alston Rivers.
Miller, D. 1991a. "South African research on Krill" *South African Journal of Antarctic Research* 21(2): 143.
Miller, D. 1991b. "Conservation of Antarctic marine living resources: a review and South African perspective" *South African Journal of Antarctic Research* 21(2): 130–42.
Miller, D. 1991c. "Exploitation of Antarctic marine living resources: a brief history and a possible approach to managing the krill fishery" *South African Journal of Marine Research* 10: 321–39.
Miller, T. ed. 1977. *Australian Foreign Minister: The Diaries of R G Casey 1951–1960* London: Collins.
Mills, G. ed. 1994. *From Pariah to Participant: South Africa's Evolving Foreign Relations 1990–1994.* Johannesburg: South African Institute of International Affairs.
Mills, G. 1996. *South African–Latin American Maritime Co-operation: Towards a South Atlantic Rim Community?* Paper presented to the Annual Navy Conference, Cape Town 17 October 1996.
Mitchell, B. 1980. "Antarctica: the unanswered question" *ICUN* 15: 9–10.
Mitchell, B. and Tinker, J. 1980. *Antarctica and its Resources* London, Earthscan.
Mohan, C. 1984. "The Antarctic regime – conflict and change at the frozen frontier" *IDSA Journal* 16: 149–70.
Molinari, A. 1988. "Refexiones sobre el funcionamiento del sistema del Tratado Antarctico" *Antartida* 17(1): 37–42.
Monmonier, M. 1991. *How to Lie with Maps* Chicago: University of Chicago Press.
Montalva, C. 1948. "Reflexiones geopoliticas sobre el presente y el futuro de America y de Chile" *Revista Geografia de Chile* 1: 27–40.
Moorcroft, P. 1981. *Africa's Superpower* Johannesburg: Collins.
Moreno, J. 1960. *Historia de las Relaciones Exteriores Argentinas 1810–1955* Buenos Aires: Editorial del Belgrano.
Morgan, J. 1996. *Australia's Antarctic Policy: Theory and Practice* Graduate Diploma in Antarctic and Southern Ocean Studies Hobart, University of Tasmania.
Morris, A. 1995. *South America: A Changing Continent* London: Hodder and Stoughton.
Morris, A. 1996. "Geopolitics in South America" in D. Preston ed. *Latin American Development: A Geographical Perspective* London: Longman: 272–95.
Morris, E. (1989a) "Sea power – South Africa between Latin America and India" *UNISA* 5: 49–51.
Morris, E. 1989b. *The Strait of Magellan* Dordrecht: Nijhoff.
Myhre, J. 1986. *The ATS: Politics, Law and Diplomacy* Boulder: Westview.
Narayana, R. and Priya, V. 1986. "Antarctica and ABC countries" in R. Sharma ed. *Growing Focus on Antarctica* New Delhi: Rajesh Publications.
Nawaz, S. 1986. "Antarctica and India's interests" *Regional Studies* 4: 65–78.
New Zealand Institute of International Affairs. 1983. *The Antarctic: Preferred Choices, Constraints and Choices* Proceedings of a Seminar held at the NZIIA, 17–18 June 1983.

Nierop, T. 1994. *Systems and Regions in Global Politics* London: John Wiley.
Nossal, K. 1989. *The Politics of Canadian Foreign Policy* Scarborough: Ontario Press.
O'Brien, R. 1992. *The End of Geography* London: Routledge.
O'Loughlin, J. ed. 1994. *Dictionary of Geopolitics* Westport: Greenwood Press.
Ommanney, F. 1938. *South Latitude* London: Longman.
Orlove, B. 1993. "Putting race in its place: order in colonial and post-colonial Peruvian geography" *Social Research* 60(2): 301–36.
Orrego Vicuna, F. 1988. *Antarctic Mineral Exploitation* Cambridge: Cambridge University Press.
Orrego Vicuna, F. 1990. "The implementation of CCAMLR: is the decision making machinery conducive to good management?" *International Challenges* 10(1): 9–20.
Orrego Vicuna, F. 1994. *The Protocol on Environmental Protection to the Antarctic Treaty: the Question of Effectiveness* IARP Publication Series Number 2 Olso: Fridtjof Nansen Institute.
Orrego Vicuna, F. and Araya, A. eds 1977. *Desarollo de la Antartica* Santiago de Chile: Editorial Universitaria.
Osherenko, G. and Young, O. 1989. *The Age of the Arctic: Hot Conflicts and Cold Realities* Cambridge: Cambridge University Press.
Ó Tuathail, G. 1989. *Critical Geopolitics: The Social Construction of Space and Place in the Practice of Statecraft* Unpublished PhD thesis, University of Syracuse.
Ó Tuathail, G. 1992. "Foreign policy and the hyperreal: the Reagan administration and the scripting of South Africa" in T. Barnes and J. Duncan, eds *Writing Worlds* London: Routledge: 155–75.
Ó Tuathail, G. 1996. *Critical Geopolitics* Minneapolis: University of Minnesota Press.
Ó Tuathail, G. and Agnew, J. 1992. "Geopolitics and foreign policy: practical eopolitical reasoning in American foreign policy" *Political Geography* 11: 190–204.
Ó Tuathail, G. and Dalby, S. 1994. "Editorial: critical geopolitics: unfolding spaces for thought in geography and global politics" *Environment and Planning D: Society and Space* 12: 513–14.
Paiva, A. 1992. *Hazanas y Rivalidades en el Polo Sur* Buenos Aires: Direccion Nacional del Antartico.
Palermo, V. 1980. "Descongelamiento de la Antartida" *Revista de la Escuela de Defensa Nacional* 27: 12–36.
Paltridge, G. and Green, J. 1995. "Antarctica and the Madrid Protocol" *Environmental Backgrounder* Number 23 Canberra: Institute of Public Affairs.
Pardo, A. 1967. Speech to the First Committee of the United Nations General Assembly 1 November 1967. New York: United Nations General Assembly.
Parker, G. 1985. *Western Geopolitical Thought in the Twentieth Century* London: Croom Helm.
Parkinson, F. 1985. "Latin America and the Antarctic: an exclusive club" *Journal of Latin American Studies* 17: 433–51.
Parsons, A. 1987. *Antarctica: The Next Decade* Cambridge: Cambridge University Press.
Peat, N. 1983. *Looking South: New Zealand Antarctic Society's First Fifty Years 1933–1983* Wellington: New Zealand Antarctic Society.
Perón, J. 1948. *Perón Expounds His Doctrine* New York: AMS Press.
Peterson, M. 1980. "Antarctica: the last great land rush on earth" *International Organization* 34: 377–403.
Peterson, M. 1988. *Managing the Frozen South: The Creation and Evolution of the Antarctic Treaty System* Berkeley: University of California Press.

Pilger, J. 1992. *A Secret Country* London: Verso.
Pinochet, A. 1984. *Geopolitica* Santiago: Editorial Andres Bello.
Pinochet de la Barra, O. 1955. *Chilean Sovereignty in Antarctica* Santiago: Editorial Del Pacifico SA.
Pinochet de la Barra, O. 1976. *La Antarctica Chilena* Santiago: Andres Bello.
Pinochet de la Barra, O. 1991. "Reminiscences of the 1959 Antarctic Treaty Conference" *Environmental Policy and Law* 21: 205–7.
Pinochet de la Barra, O. 1995. "The development of the Antarctic Treaty System: review of past developments" in A. Jackson ed. *On the Antarctic Horizon: Proceedings of the International Symposium on the Future of the Antarctic Treaty System* Hobart: Australian Antarctic Division: 7–11.
Pittman, H. 1981. *Geopolitics in the ABC Countries* Unpublished PhD thesis Washington DC, The American University.
Pittman, H. 1988a. "El impacto geopolitico sobre la democratization y conflictos en el Cono Sur" *Geopolitica* 37: 11–52.
Pittman, H. 1988b. "From O'Higgins to Pinochet: Applied geopolitics in Chile" in P. Kelly and J. Child eds *Geopolitics of the Southern Cone* Boulder: Lynne Rienner: 173–86.
Plott, B. 1969. *The Development of US Antarctic Policy* Unpublished PhD thesis Fletcher School of Diplomacy Tufts, University of Boston.
Pratt, M. 1992. *Imperial Eyes* London: Routledge.
Price, A. 1962. *The Winning of Australian Antarctica* Sydney, Heinemann.
Pringle, T. 1991. "Cold comfort: the polar landscape in English and American popular culture" *Landscape Research* 16: 43–8.
Prior, S. 1997. *Antarctica: View from a Gateway* Wellington: Centre for Strategic Studies Victoria University of Wellington.
Puri, M. 1986. "Geopolitics in the Indian Ocean – the Antarctic dimension" *International Studies* 23: 155–68.
Quartermain, L. 1952. "Gateway to the South" in F. Simpson ed. *The Antarctic Today* Wellington: AH and AW Read: 345–64.
Quartermain, L. 1967. *South to the Pole* London: Oxford University Press.
Quartermain, L. 1971. *New Zealand and the Antarctic* Wellington: Government Printers.
Quartermain, L. 1981. *Antarctica's Forgotten Men* Wellington: Millwood Press.
Quigg, P. 1983. *A Pole Apart?* New York: McGraw-Hill.
Quilty, P. 1990. "Antarctica as a continent for science" in R. Herr, R. Hall and B. Davis eds *Antarctica's Future: Continuity or Change?* Canberra: Australian Institute of International Affairs: 29–37.
Rabbani, M. 1994. "Pakistan's interest in the Antarctic" in R. Herr and B. Davis eds *Asia in Antarctica* Canberra: Centre for Resource and Environmental Studies (ANU) 119–22.
Radcliffe, S. 1996. "Imaginative geographies, post-colonialism and national identities: contemporary discourses of the nation in Ecuador" *Ecumene* 3(1): 5–21.
Radcliffe, S. and Westwood, S. 1996. *Remaking the Nation: Place, Identity and Politics in Latin America* London: Routledge.
Reboratti, C. 1982. "Human geography in Latin America" *Progress in Human Geography* 6: 397–406.
Reboratti, C. 1983. "El encanto de la oscuridad: notas acerca de la geopolitica en la Argentina" *Desarrollo Economico* 23: 137–44.
Riesco, J. 1985. "Geopolitica austral y Antarctico" *Boletin Antarctio Chileno* 4: 14–17.
Riffenburgh, B. 1994. *The Myth of the Explorer* London: Belhaven.

Rock, D. 1987. *Argentina 1516–1897* London: I.B. Tauris.
Rojas, A. 1994. "The environmental movement and the environmentally concerned scientific community in Chile" *European Review of Latin American and Caribbean Studies* 56: 93–118.
Rothwell, D. 1992. *The Madrid Protocol and its Relationship to the ATS* Antarctic and Southern Ocean Law and Policy Occasional Papers number 5, University of Tasmania.
Rothwell, D. 1995. "Polar lessons for an Arctic regime" *Cooperation and Conflict* 29: 55–76.
Roucek, J. 1981. "Geopolitics in Antarctica and the Falkland Islands" *World Affairs Interpreter* 22(1): 44–56.
Routledge, P. 1996. "Critical geopolitics and terrains of resistance" *Political Geography* 15: 509–32.
Rowlands, I. 1996. "South Africa and global climate change" *Journal of Modern African Studies* 34: 163–78.
Russell, R. 1988. "Argentina: Ten years of foreign policy towards the Southern Cone" in P. Kelly and J. Child eds *Geopolitics of the Southern Cone and Antarctica* Boulder: Lynne Rienner: 67–82.
Said, E. 1993. *Culture and Imperialism* London: Chatto and Windus.
Scholes, R. 1954. *Seventh Continent* London: George Allen and Unwin.
Schroff, G. 1982. "Antarctica: politics and resources" in A. Peers ed. *Seaford House Papers* Sandhurst: Royal Military College: 99–125.
Scilingo, A. 1973. *El Tratado Antarctico: Defensa de la Soberania y la Proscipcion Nuclear* Buenos Aires: Liberia Hachette.
Scully, T. and Kimball, L. 1989. "Antarctica: is there life after minerals?" *Marine Policy* 13(2): 87–98.
Shackleton, E. Lord. 1985. "Why the Falklands Matter" *The Times* 22 April 1985.
Shapiro, M. 1991. "Sovereignty and exchange in the orders of modernity" *Alternatives* 16: 447–77.
Shapiro, M. 1994. "Moral geographies and the ethics of post-sovereignty" *Public Culture* 6: 479–502.
Shapiro, M. 1988. *The Politics of Representation* Madison: University of Wisconsin Press.
Shapley, D. 1983. "International treaty still on ice" *Nature*: 301–63.
Shapley, D. 1985. *Seventh Continent* Washington DC: Resources for the Future Inc.
Sharp, J. 1993. "Publishing American identity: popular geopolitics, myth and the *Reader's Digest*" *Political Geography* 12: 491–503.
Sharp, J. 1995. "Desperately seeking a new world order: The *Reader's Digest* at the end of the Cold War" Paper presented at the "Geopolitical reconstruction and the new world order" conference, Royal Holloway, University of London 5 May 1996.
Shumway, N. 1991. *The Invention of Argentina* Berkeley: University of California Press.
Shusterich, K. 1984. "The Antarctic Treaty System: history, substance and speculation" *International Journal* 34: 800–11.
Sidaway, J. 1994. "Geopolitics, geography and terrorism in the Middle East" *Environment and Planning D: Society and Space* 12.
Sidaway, J. and Simon, D. 1993. "Geopolitical transition and state formation: the changing political geographies of Angola, Mozambique and Namibia" *Journal of Southern African Studies* 19: 6–28.
Siegfried, W. 1985. "Krill: the last unexploited food resource" *Optima* 33: 67–79.

Siegfried, W. 1991. "Three decades of South African science in Antarctica" *South African Journal of Antarctic Research* 21: 225–9.
Simon, D. 1991. "The ties that bind: decolonization and neo-colonialism in southern Africa" in C. Dixon and M. Heffernan eds *Colonialism and Development in the Contemporary World* London: Mansell: 21–44.
Simpson, F. 1954. "New Zealand and Antarctica" *New Zealand Geographer* 10(1): 1–24.
Simpson-Housley, P. 1992. *Antarctica: Exploration, Perception and Metaphor* London: Routledge.
Smith, G. 1994. "Political theory and human geography" in D. Gregory, R. Martin and G. Smith eds *Human Geography* London: Macmillan: 54–77.
Smith, N. and Godwelska, A. eds 1994. *Geography and Empire* Oxford: Blackwell.
Smuts, J. 1925. "Foreword" in C. Weidner *Antarctica's Influence on our Rainfall* South African Association for the Advancement of Science, Good House, July 1925.
South Commission. 1987. *The Challenge to the South* Oxford: Oxford University Press.
Speake, P. 1996. *The Scottish National Antarctic Expedition: nationalism, science and idealism* Paper presented to the Annual Conference of the Institute of British Geographers 3–6 January 1996.
Spence, J. 1983. "The Falklands conflict" *Optima* 31(3): 107–20.
Sreenivasan, T. 1994. "India's Antarctic policy and interests" in R. Herr and B. Davis eds *Asia in Antarctica* Hobart: Government of Tasmania: 73–8.
Stokke, O. and Vidas, D. eds 1996. *Governing Antarctica* Cambridge: Cambridge University Press.
Storni, S. 1967. *Intereses Argentinos en el Mar* Buenos Aires: Centrol Naval, Instituto de Publicaciones Navales.
Sugden, D. 1982. *Arctic and Antarctic* Oxford, Blackwell.
Sugden, D. 1989. "The polar environment: illusion and reality" *Ambio* 18: 2–5.
Sullivan, W. 1957. *Quest for a Continent* London: Secker and Warburg.
Sullivan, W. 1961. *Assault on the Unknown* New York: Secker and Warburg.
Suter, K. 1980. *World Law and the Last Wilderness* Sydney: Pluto Press.
Suter, K. 1991. *Antarctica: Private Property or Public Heritage* London: Pluto.
Swan, R. 1961. *Australia in Antarctica* Melbourne: University of Melbourne Press.
Taubenfield, H. 1961. "A treaty for Antarctica" *International Conciliation* January: 249–55.
Taylor, G. 1920. *Antarctic Adventure* London: George Allen and Unwin.
Taylor, P. 1993. *Political Geography* London: Longman.
Toledo, X. and Zapater, E. 1991. *Geografia General y Regional Chile* Santiago: Editorial Universitaria.
Toma, P. 1956. "Soviet attitudes towards the acquisition of territorial sovereignty in the Antarctic" *American Journal of International Law* 17: 195–228.
Tonnessen, J. and Johnson, A. 1982. *The History of Modern Whaling* London: C Hurst.
Triggs, G. 1986. *International Law and Australian Sovereignty in Antarctica* Sydney: Legal Books.
Unger, R. 1991. *The Limits to Learning: An examination of the influence of consensual knowledge and epistemic communities in international relations* Unpublished research essay Norman Paterson School of International Affairs, Carleton University.
Vale, P. 1996. "Zuid-Afrika's zoekocht naar een buitenlands beleid" *Internacionale Spectator* 50: 243–9.

Venter, D. 1996. "South Africa, Brazil and South Atlantic security: towards a zone of peace and cooperation in the South Atlantic" Paper presented to the International Relations Research Institute on Brasil-Africa du Sol, Rio de Janeiro 24–5 September 1996.
Viall, J. 1991. "South Africa: the road to the Antarctic Treaty" *South African Journal of Antarctic Research* 21(2): 125–9.
Vidas, D. 1992. *Antarctic Tourism: A Challenge to the Legitimacy of the Antarctic Treaty System?* IARP Publication Series Oslo: Fridtjof Nansen Institute.
Vidas, D. 1994. *The Antarctic Treaty System in the International Community: A Generalised Assessment* IARP Publication Series Number 1 Olso: Fridtjof Nansen Institiute.
Vogler, J. 1995. *Global Commons* Chichester: John Wiley.
Walker, R.B.J. 1988. *One World, Many Worlds: Struggles for a Just World Peace* Boulder: Lynne Rienner.
Walker, R.B.J. 1993. *Inside/Outside* Cambridge: Cambridge University Press.
Wallerstein, I. 1983. *Historical Capitalism* London: Verso.
Watts, A. 1992. *International Law and the Antarctic Treaty System* Cambridge: Cambridge Unversity Press.
Wight, M. 1995. *Power Politics* Leicester: Leicester University Press.
Willetts, P. 1978. *The Non-Aligned Movement* London: Pinter.
Winnichakul, T. 1994. *Siam Mapped: A History of the Geo-body of a Nation* Honolulu: University of Hawaii Press.
Wohl, R. 1994. *A Passion for Wings: Aviation and the Western Imagination 1908–1918* New Haven: Yale University Press.
Wolf-Philips, L. 1987. "Why Third World? Origins, definition and usuage" *Third World Quarterly* 9(4): 1311–28.
Wolfrum, R. 1995. "Possible challenges and the future development of the Antarctic Treaty System" in A. Jackson ed. *On the Antarctic Horizon: Proceedings of the International Symposium on the Future of the Antarctic Treaty System* Hobart: Australian Antarctic Division: 85–91.
Woolcott, R. 1984. "The Future of the Antarctic Treaty System" in L. Alexander and L. Carter eds *Antarctic Politics and Marine Resources* Kingston, Rhode Island: Center for Ocean Management Studies.
Woolcott, R. 1990. Speech to the Australian Institute of International Affairs 10 August 1990 *World Review* 30: 48–56.
Yahuda, M. 1996. *The International Politics of Asia-Pacific 1945–1995* London: Routledge.
Zegers, F. 1983. "The Canberra Convention: objectives and political aspects of its regulation" in F. Vicuna ed. *Antarctic Resources Policy* Cambridge, Cambridge University Press.

Index